Learning and Teaching with Geomedia

Learning and Teaching with Geomedia

Edited by

Thomas Jekel, Eric Sanchez, Inga Gryl,
Caroline Juneau-Sion and John Lyon

CAMBRIDGE
SCHOLARS
PUBLISHING

Learning and Teaching with Geomedia,
Edited by Thomas Jekel, Eric Sanchez, Inga Gryl, Caroline Juneau-Sion and John Lyon

This book first published 2014

Cambridge Scholars Publishing

12 Back Chapman Street, Newcastle upon Tyne, NE6 2XX, UK

British Library Cataloguing in Publication Data
A catalogue record for this book is available from the British Library

ISBN (10): 1-4438-6213-4, ISBN (13): 978-1-4438-6213-4

Work on this publication has been co-funded by the LLP-Comenius network digital-earth.eu (510010-LLP-1-2010-1-AT -COMENIUS-CNW) funded by the European Commission:

TABLE OF CONTENTS

PREFACE

Our lives are thoroughly 'geomedialized'. Geomedia technologies that emerged and – particularly important – were popularized in the course of the last decade shape the opportunities we have to appropriate the world around us and to act within society. Global positioning systems, ubiquitous computing, web mapping in connection with remote sensing and the convergence of media contribute to a geoweb providing an increasingly powerful aspect of our everyday action in space. The use of these technologies is a very specific aspect of an emerging digital culture that our schools have to prepare for.

We receive news such as twitter messages, 'just in time, geo-localized' on digital globes. Web information is available everywhere, as is the cloud. Our lives can be recorded within social networks, visualized in a framework of time and space. We are guided in both the countryside and cities by GIS navigation systems. Our view of the city is extended by applications combining geo-localization, image recognition and web content, adding meaning to the things we see. The smartphone alerts us when we are near to certain sites, and spatial referencing offers the possibility of combining personal interests with spatial information. Apps synchronize geo-location with our personal timetable on a tablet providing the latest traffic and transport information for our current position. We know where our social network friends are and can arrange spontaneous (face-to-face) meetings. We produce spatial data by evaluating a restaurant we have visited, or simply, even without realizing it, by having been in an area while looking up a certain word in a search engine. Each photo taken with data glasses is uploaded while being enriched with various metadata. Finally, we can produce and design our own maps with user-friendly web mapping tools. In the end, we may encounter mixed realities enriched by digital and geolocalized data and a merging of real and virtual spaces.

In sum, we constantly consume and produce media which carries geo-referenced information. In the context of this book, we refer to these kinds of media as 'geomedia'. Geomedia are not only cartographic media but also geo-tagged pictures and written descriptions of paths and places. Today, geomedia come with an enormous semantic field closely linked to web2.0, the Internet of social interaction. To be more concise, geomedia connects the web2.0 with the lifeworld by providing a reference to physical human beings. The 'virtual world' of the web seems to belong to yesterday; the web of today clearly enriches our everyday lives. A renaissance of space and place foils the idea of overcoming space in a digital age. At the same time, geomedia and the resulting data sets are media of surveillance and control, of intrusion into intimacy and the use of very private spatial data for commercial aims, and these clearly change our concepts of privacy and culture as a whole. To conclude, geomedia are ever-present in everyday life; they shape action and encourage thinking and reasoning – and their use needs reflection and reflexivity.

Education has to enable students and teachers to come to terms with this development. This carries two main aims: learning to act with geomedia in everyday contexts, and utilizing geomedia to understand spatial problems in the domain of science. In other words, the geomedia society requires abilities and capabilities to utilize geomedia, as they are a central medium for learning, thinking and communicating about spaces, as well as for achieving and retaining power and influence over places. Participation in society therefore increasingly needs geomedia literacy. The task for geomedia education is, in a minor way, to teach technical skills; in a major way, it is to enable the mature and self-directed utilization of geomedia.

The aim of this book, therefore, is to support teachers and teacher trainers, through an overview of current technical, social and pedagogical approaches and through ready-to-use learning environments. The collection of learning environments utilizes geomedia to support learning and problem solving in the spatial domain. The book offers new ideas for implementing teaching strategies with geomedia that foster students' creativity and participatory abilities. It contributes to the education and empowerment of spatial citizens who are able to act in a complex, geomedialized world, while being adapted to it, but also constructing, shaping and reinventing it.

This book is an outcome of the digital-earth.eu project, an EU Comenius Multilateral Network that aims at the promotion of geomedia usage and teaching in schools. It is congruent with the network's purposes to share innovative best-practice examples to inspire and encourage transfer beyond national borders. This book unites contributions from researchers and practitioners from all over Europe. All learning environments are field-tested and transferable and provide the basic considerations and materials to be adapted by the reader.

The structure of this book is given by placing the learning environments (Part II) in a kind of continuum. It starts with simple technical aspects of mapping, and introduces the challenges of spatial thinking. These learning environments support consumption as well as map production abilities and capabilities, encompassing single-step analytical tasks leading to hypothesis construction, as well as exercises to collect, select and modify spatial data, to comment on visualizations and to design one's own. Further papers in this book address the necessity for the reflective and reflexive utilization of geomedia. Reflection refers to the deconstruction of societal discourses hidden in geomedia, a change of perspective that allows the evaluation of a geomedium's ability to handle a certain task or problem. Reflexivity relates to the capacity to produce conscious hypotheses by being aware of one's own construction of spatial scenarios based on medium, preconditions, and one's own interests, which prepares for conscious, mature and responsible communication with geomedia.

The book concludes with learning environments aiming at participation in society through geomedia. These environments refer to the expression of interests and broad communication and negotiation using geomedia in a web2.0 context and address citizenship competences, as described in Spatial Citizenship.

This order is not necessarily an order of complexity: each learning environment may be used in, or transferred to, different levels of achievement.

The book is framed by three introductory chapters to contextualize the learning environments (Part I). The first envisions the rapid technological development of geomedia; the second discusses the influence of geomedia on society, societal power-relations and the consequences for everyday action and decision making; the third discusses and illustrates the pedagogical background of learning with and about geomedia.

The editors hope that the book will help teachers to use the learning environments, adapt them to their classroom situations, and enhance them or build similar learning environments for their own teaching. The book should inspire teachers to experiment and creatively develop similar or altogether different learning environments, helping students to navigate more competently the emerging geoinformation society. The technology is there and easy to use, so teachers and teacher trainers are encouraged to contribute to geomedia education through their own approaches.

The editors strongly believe that this endeavour is not simply about adaptation to a new technology and its associated culture. We suggest that geomedia are helpful for our students to re-invent their world and society; they are not just a means of handling the given world. Geomedia in this case become a medium of empowerment for young people to actively design their worlds.

The editors are indebted first to the authors who have borne with them through delays in the process. All of the papers underwent several rounds of review, and even so, we could not publish all papers on offer. Beyond the editorial team, Nicole Ferber (Salzburg) has been incredibly helpful with all institutional issues at the coordinating institution of the digital-earth.eu network, as well as with the final version of the book. John Lyon did proofreading despite being always busy at the Geographical Association, Sheffield, UK. Mary Rigby (Glasgow) was amazingly helpful doing copy-editing. Viola Geiger (Salzburg) expertly did the final formatting with help from various staff at Cambridge Scholars Publishing.

During work on this book, four digital earth natives were born to members of the editorial team in late 2012 and early 2013, enlivening editorial meetings, and both hampering and inspiring our work: Camille, Anouk, Eleanor and Hannah.

We would like to dedicate the book to them.

PART ONE:

GISOCIETY AND GEOMEDIA EDUCATION

CHAPTER ONE

TECHNOLOGICAL FOUNDATIONS FOR THE GISOCIETY

JOSEF STROBL

Abstract

Based on the observation that today's 'information society' is based increasingly on leveraging the added value of 'location', this chapter briefly explores some of the main technological foundations for this trend. It is demonstrated that a set of geospatial techniques, as well as general ICT trends highly relevant for location-based applications, are facilitating and sometimes driving the emergence of a highly spatially aware digital practice of personal, communal and societal interaction. These developments generate questions regarding required literacies, which should be considered essential components in teaching and learning processes.

1 Introduction

Society today is frequently labelled an 'information society' in an age of pervasive digital technologies. Megatrends are driving current developments. One of these is ubiquitous and permanent connectivity. Being 'always on (line)' was expected to make people less dependent on the operational contexts of a location, as was speculated in the 'Death of Distance' debate (*Economist*, 1995; Cairncross, 1997). Just as the Internet originally had a military background, the use of satellite signals for positioning triggered another megatrend – one of pervasive positioning.

As this technology moved into consumer electronics, it triggered the 'Revenge of Geography' discourse (*Economist*, 2003). Obviously, the information society did not turn its citizens into location agnostics, but rather led to the contrary, the dominance of 'where'. Perhaps triggered by the uncoupling of communication from co-presence, bringing location into the personal digital realm was urgently needed to balance the effects of telecommunication.

Thus the 'spatial turn' into a Geo-Information Society is obvious throughout its digital infrastructure, as well as in its daily workings. Very few tasks, interests and activities of a person are independent of location. Location is not only a measurable position; it defines what is around it by proximity and distance. Location very effectively serves as an integrator across all domains – database experts would consider position as a key facilitating 'spatial join' operation. This spatial integration across 'layers of information' or domains of interest effectively turns locations into places (Goodchild, 2011).

With the society's realm defined by spaces, and people invariably associated with places, the concept of a Geo-Information Society gains solid traction. To be a citizen in this society, a common set of knowledge, skills and qualifications is needed. Literacy and numeracy historically have been requirements for participation in a community. Digital literacy is generally accepted as the passport into the information society, and from a GISociety perspective a certain set of geospatial literacy elements will be required.

This can be loosely connected with the traditional map-reading skills which are required as a foundation to build spatial orientation and to complete certain spatial tasks such as navigation. From a somewhat broader perspective, an individual's general spatial awareness is the foundation for spatial thinking, which supports citizens in their participation in society.

Which competences and skills are required and which will (partially) replace traditional map reading and interpretation? While this question will be dealt with in subsequent chapters, it is clear that we are aiming for a convergence of education demand and technology supply: new technologies will emerge and offer opportunities, while only those which address a (maybe new) demand will be sustained in the longer term. Technology of course is not an end in itself, but it has in the past changed societies, and it will do so in the future.

From this perspective, several key geospatial technologies are explored regarding their impact on individual action and communication patterns, and thus on participation in societal behaviour. This can only be a snapshot of current developments, and an individual assessment of relevance. Nonetheless, any such *tour d'horizon* of geospatial technologies will demonstrate the dynamics of this field and the importance of considering its links with education.

2 Key Technologies

Innovation, be this in knowledge or in technologies, has always been a driving force in expanding the horizons of humankind – literally and metaphorically. Navigating and exploring the surface of our globe depended, and still depends, on the navigation technologies available to travellers. Enjoying and leveraging the benefits of a bird's-eye perspective require flight, which opened up the huge potentials of remote sensing. Societies worldwide today share far more similar traits than did the societies of our ancestors, based on communication technology and media reaching across the globe.

Today we use the term 'Geo-Information Society' partly to express the understanding that 'GIS' is not primarily a set of technologies, but more importantly creates societal dynamics around the concepts of place and space. Still, to understand and to explore a GI Society we need to look at (some of) the current technical developments facilitating and sometimes driving progress towards a spatially 'smart' society. What are the 'top 7' geospatial technologies most relevant for the general public and thus society at large?

Universal positioning. Knowing where we are at any given moment defines the starting point for navigation, defines distance and proximity, establishes a personal 'environment', and offers choices of what to do, who to meet and where to move. Easy access to Global Navigation Satellite Systems (GNSS) signals like GPS plays a central role, particularly as most of our personal devices such as smartphones and tablets work as receivers. The resulting 'pervasive positioning' of people, vehicles and other assets is at the core of permanent spatial awareness, using the idea of *where* to learn about *what* and *who*. It therefore comes as no surprise that technologists work hard to overcome the limitations of GNSS inside buildings, in urban canyons and underground. A range of solutions have been proposed, and some are being implemented, to keep track of location outside of the range of satellite signals: WiFi positioning and dead reckoning via acceleration are just two of the approaches currently addressing the need for pervasive positioning.

Personal sensors. Beyond positioning, the latest personal devices contain a growing number of sensors. The accelerometer in each smartphone is complemented by a digital compass and general attitude sensor, and other sensors that take readings for sound and light. Our personal devices are increasingly powerful because they serve as sensor integration platforms with a strong communication back end. This in turn facilitates a context-enriched filtering of information: when driving fast, less local detail is shown on a map. When it is dark, different navigational cues are offered.

Location-enabled apps. When looking for a place to stay, a hotel booking app explores 'nearby' first. Other searches default to a 'minimum distance' criterion. The first law (or, rather, rule) of GISocietal applications is to 'act locally'; starting from the assumption that individual context defines the area of (potential) action (AoA). We will be assisted by many more, and sometimes perhaps rather obtrusive apps, using location to connect information from different sources – for better or worse.

Augmented reality (AR). At a 'local scale', within range of our visual information, data about what we can see is being directly fused into our visual channel of perception. This idea has been pioneered by heads-up displays on the flight decks of aeroplanes, and is now being taken to mass markets through Google's glasses. Every smartphone or tablet can serve as an AR device by switching on the camera, and annotating what we see. This might be helpful for tourists, but might perhaps be even more so for locals, replacing the rather cumbersome map interface in personal navigation.

Perspective view. Traditional maps, as 'flattened views', require significant abstraction of spatial orientation and message. Interestingly, we are now moving (back?) from the 'neutral', 'objective' presentation of spatial data in the same fashion for everyone – as on a standardized map – towards a subjective, individualized perspective, where everyone controls their own view. Mapping and cartography are evolving from a product-centric view into a communication science emphasizing individualized user interfaces. The observation point in a perspective view can be a person's current location (in this case, see AR), or is actively controlled to explore an area of interest from different angles. The reasons for the attractiveness of Google Earth's and similar interfaces are twofold: less abstraction, and more control for the viewer. Perspectives are closer to the everyday experience, make information more accessible. Thus maps are increasingly replaced with perspective views, even though some areas are obscured, scale changes with depth, and everyone's perspective is different: this all makes for a more personalized experience.

Click to touch to kinetics. Interaction paradigms change, largely because of technological developments. Not to start at the digital dinosaur age of typing-based interaction, the mouse as a point-and-click device used to be the defining means of human–computer interaction. As computers shrink into mobile personal devices which are available in any kind of context, there is no space, and no use for an indirect interaction tool like a mouse. We point and touch with our fingers, giving new meaning to the term 'digital' technology. Today's generation growing up as digital natives will use a mouse-like pointer when 'working' on a desktop, but not for the pervasive interaction in all kinds of mundane tasks. 'The reach of touch', though, is limited – by the size of a screen and the need to be in physical contact. Kinetic interaction through physical gestures is now transitioning from games to more general contexts, generating interesting research questions as to which gestures will emerge as standardized spatial actions, e.g. corresponding to the now familiar pincer movement to change scale on a touch pad. Where do we go from kinetics? Well, research demonstrates the feasibility of mental control. But most of our thoughts are in specific languages. Thus it might be easier to stick with touch and motion for now.

Personal and inter-personal devices. Similar to the emergence of the personal computer some thirty years ago, another big step has led to a personal device in everyone's pocket. Computer classrooms are already on their way into the dustbin of history, as there is no need for pupils to go to a computer any more since devices are always with them. Meetings and presentations are labelled 'BYOD' ('bring your own device'), to facilitate digital interaction and tools. Watching many individuals today demonstrates that personal devices are considered an integral extension of one's self and that the majority of interactions, and perhaps of learning, occur through digital interfaces (Strobl, 2007), even when there would be alternative 'face-to-face' means of communicating. This development has come about thanks to personal devices serving as very effective, multi-functional communication assistants (remember, some early smartphones were branded as 'communicators'). Frequently, the main purpose is communication; instead of, originally, being only marginally connected with computers. Much of what serves the purpose of connecting people and interests via telecommunication, the lightweight personal character of devices, plus direct-touch interaction obviously facilitate collaborative action. Interacting jointly in front of one display is certainly done more now than when the holder of the mouse was in exclusive control.

Many more technical developments could be listed here as driving forces for change, for change in individuals' actions, change in how communities and societies work, and of course change in the skills and qualifications to be considered in education. One example of a rather invisible development at the back-end of digital interfaces discussed above is the emergence of cloud computing. Having originally started with email and now expanding into virtually all areas of computing, only a web browser or a dedicated app is needed to do the job. All storage and functionality, plus the authentication required for access, are available online 'on the Internet'. Users tare not tied to just one or a few computers where 'their' data and software are installed; essentially any kind of Internet connection is sufficient.

This paradigm change in computing has far-reaching consequences, and perhaps the biggest game-changer of all is the perspective of 'everybody' using an impressive array of digital technologies. Much lower levels of computing skills are needed, and end users can focus more on what to do and less on how to do things. The need to manage one's own computer has been removed, personal devices serving as access points to cloud services which update themselves and are easily interchangeable.

3 Technology Education?

Exploring these and other current and emerging technologies relevant from a geospatial viewpoint gives rise to questions of educational objectives. While technology education is a value in and of itself (as outlined in NRC, 2006), it is not the focus of discussion here. Citizens in a GISociety require competences based on technologies, but relating to their personal lives and participation in society.

Maps, images and other sensor readings are frequently viewed as digital representations, depicting models of the real world. This virtual world helps with monitoring, understanding and managing the real world. While this concept of a duality of the real world and its virtual representation (Strobl, 2009) is helpful in typical GIS applications, it falls short when exploring the role of geospatial technology brought into individuals' lives through personal devices.

We increasingly observe a fusion of real and virtual worlds in people's lives. Digital communication and action create realities, and a virtual representation might be more relevant than its physical counterpart. The separation of the real world and its virtual model is giving way to an integrated entity, which is perhaps best conceptualized as a 'Digital Earth' framework where digital interfaces define and facilitate the interaction patterns among people (Fischer, 2008), between people and their environments, and are constituent for a GISociety.

Such a 'Digital Earth' (Craglia et al., 2008) clearly generates new demands on competences and skills. Many of these are explored in this book. Beyond that, education needs to explore the additional risks and dangers associated with emerging technologies. Location is an extremely powerful factor in privacy considerations, precisely because it allows the joining of small bits of information. Therefore a critical assessment of possible violations needs to be an integral part of learning. While opting in via a disclose-my-location-for-my-current-task button would be desirable, frequently not even an opt-out option is granted. We have seen anecdotal evidence of the potential of face recognition for tracking individuals, even without using a personal device: awareness of the risks associated with a highly connected digital society is a primary aim of all education.

Participation is a broadly used buzzword in many domains. The step from the Internet as just another media pathway towards 'web2.0' where the separation of sender and recipient, of author and reader, is giving way to a prosumer who can swap the roles of creator and user of information anytime is also taking place in the geospatial context (Scharl & Tochtermann, 2007). Increasingly a passive, 'consumerist', role is being replaced by a more active participation in a multi-way communication culture of 'likes', ratings and comments. Participating in a GISociety starts with these established actions, but expands towards contributions to OpenStreetMap (Ramm & Topf, 2009) as the 'Geographers' Wikipedia', the uploading of a hiking track to a community website, and the volunteering of skills and expertise for a regional portal. However, contributions of geospatial information are currently both heavily gendered (i.e., almost exclusively male) and limited to educated strata of society (Steinmann, Häusler, Klettner et al., 2013), providing ample scope for specific education strategies.

Of course the GISociety is only one of many overlapping perspectives on a rapidly evolving digital age. Yet, as people are always present in a place, interact within the spaces of their environment, take interest in regions and destinations and think about the world's dynamics, geographic information plays an indispensable role in life. Perceptions of, and thus actions in, the 'Digital Earth' are greatly influenced by the geospatial technologies available. Teaching and learning with these is, and will be, a stimulating challenge.

References

Anon. (2003): 'The Revenge of Geography'. *The Economist* 366 (8315), (15 March 2003), 19–22.

Cairncross, F. (1995): 'The Death of Distance'. *The Economist* 336 (7934) (30 September 1995), 16–17.

—. (1997): *The Death of Distance*. Cambridge, MA: Harvard Business School Press.

Fischer F. (2008): 'Location Based Social Media – Considering the Impact of Sharing Geographic Information on Individual Spatial Experience'. In: A. Car, G. Griesebner & J. Strobl (eds.), *Geospatial Crossroads @ GI_Forum '08. Proceedings of the Geoinformatics Forum Salzburg*. Berlin: Wichmann, 90–96.

Goodchild, M.F. (2011): 'Formalizing Place in Geographic Information Systems'. In L. M. Burton, S. P. Kemp, M.-C. Leung, S. A. Matthews & D. T. Takeuchi (eds.), *Communities, Neighborhoods, and Health: Expanding the Boundaries of Place*. New York: Springer, 21–35.

Gould, M., Craglia, M., Goodchild, M. F., Annoni, A., Camara, G., Kuhn, W., Mark, D., Masser, I., Maguire, D., Liang, S. & E. Parsons (eds.) (2008): 'Next-Generation Digital Earth'. *International Journal of Spatial Data Infrastructures Research* (3), 146–167.

NRC (2006): *Learning to Think Spatially: GIS as a Support System in the K-12 Curriculum*. Washington, DC: National Academics Press.

Ramm, F. & Topf, J. (2009): *OpenStreetMap: Die freie Weltkarte nutzen und mitgestalten*. 2nd edn. Berlin: Lehmanns Media.

Scharl A. & Tochtermann, K. (eds.) (2007): *The Geospatial Web. How Geobrowsers, Social Software and the Web 2.0 are Shaping the Network Society*. New York: Springer.

Steinmann, R., Häusler, E., Klettner, S., Schmidt, M. & Lin, Y. (2013): 'Gender Dimensions in UGC and VGI: A Desk-Based Study'. In T. Jekel, A. Car, J. Strobl & G. Griesebner (*GI_Forum 2013*), *Creating the GISociety*. Berlin/Offenbach: Wichmann & Wien: Austrian Academies of Sciences Press, 355–364.

Strobl, J. (2007): 'Geographic Learning in Social Web Environments'. In S. Catling & L. Taylor (eds.), *Changing Geographies: Innovative Curricula*. Oxford: IGU Commission for Geographical Education / Oxford Brookes University, 327–332.

—. (2009): 'Geo-ICT: Connecting Physical and Virtual Geographies'. In H. Scholten, R. van de Velde & N. van Manen (eds.), *Geospatial Technology and the Role of Location in Science*. Berlin: Springer GeoJournal Library, Vol. 96, 187–199.

CHAPTER TWO

EVERYDAY GEOMEDIA USE AND THE APPROPRIATION OF SPACE

FLORIAN FISCHER

Abstract

The geoweb (i.e. the convergence of geospatial technology, mobile web technologies and social media) has recently enabled an environment for mass participation in the production and distribution of geographic information. A vivid mash-up culture, resulting in the emergence of a variety of geomedia, is promoting space to a new paradigm for online search, communication and interaction. Geomedia provide a way to gather data concerning physical location in today's networked and multiplexed spatialities. While everyday life is shaped by physical nomadism, flexibilization, fragmentation and global flows of goods and services, geomedia facilitate new relationships between people and places far beyond a filtering by physical proximity alone. Geomedia serve users wishing to master urban heterogeneity and are used to facilitate new forms of empowerment in the appropriation of public space. Contributing users become new gatekeepers who mediate between local businesses and customers. New modes of the production of spatial meaning emerge, including interest groups with the capability of contributing to the process. Hence users of these services can be considered as agents of change in networked urbanity. We have only just begun to understand how digital citizens re-form public space in the context of its transformation by the geoweb. Based on a case study, this chapter reveals the transformative capacity of the geoweb for the 'making' of public space. It depicts the filtering, sorting and commodification of public space, and the role of different modes of collaborative mapping.

1 The Geoweb

The geoweb environment is normally associated with digital maps, routing services and locating services, but it also describes a broader phenomenon concerning the convergence of mobile communication, Internet technology and geospatial technology.

The geoweb links potentially global web-based activities to territorial snippets on the earth's surface and enables people to act upon, or with reference to, a location (be it their own, or that of an object or person). This ability is facilitated by a user-friendly environment for the geospatial organization and use of information, services and applications on the Internet.

The geoweb facilitates the democratization and increasing ubiquity of the production and distribution of geographic information on the Internet. It has disrupted a GIS-industry with its institutionalized practices for the capture, processing and provision of geospatial services for specific scientific, political and economic target groups. Two major shifts can be identified. First, a ubiquitous and user-friendly provision of geospatial services has extended the potential market for geographic information beyond GIS-experts towards a mass market of Internet users, and opened up an unforeseeable range of novel practices in everyday life. Second, the social software approach has democratized the production of geographic information, making every user into a potential contributor, allowing everyone to share, communicate through, collaborate in and produce geographic information.

2 The Democratization of Geographic Information

For a number of years, the production of geographic information has rested in the hands of private and governmental mapping agencies that spent billions to create large-scale geographic data sets, run frequent updates and provide geographical completeness. This costly process is a driver for the collection of only a limited and rigorous selection of geographic information – that is, the production of geographic information aligned to the narrow demands of certain administrations (see upper section of Figure 2-1) and enterprises or, at best, to the most popular everyday demands (e.g. navigation services), as the costs of production are very high (Fischer, 2010a). Google Maps, Bing Maps, Yahoo! Maps and not least the OpenStreetMap (OSM) have overcome these previous shortcomings. Free-of-cost geo-browsers have revolutionized the domain of consumer and citizen services, focusing on social location applications for leisure, tourism, entertainment and political participation (Fischer, 2010b).

First, they provide a graphical interface for a global coverage of map and satellite data, following a model of free mapping services: the use is broadly free, but apart from OSM, the geo-data is not freely accessible. Secondly, geobrowsers provide a central functionality that allows combination with hitherto unavailable spatial web resources, for example a geocoding service that converts an address or place name to coordinates (Purvis & Svennerberg, 2010). In combination with their free mapping services and APIs (Application Programming Interfaces), new geo-aware media (or geomedia) have emerged that link, process and visualize content from multiple sources by geographic reference in a way that is seamless to the user. These are termed mapping mash-ups (Carl, Zund & Zund, 2008). The introduction of geobrowsers is commonly considered a landmark for the growth of the geoweb environment as they supported the colonization of a mass-market segment (see lower section of Fig. 2-1) of Internet users by spatializing web services (Fischer, 2010a).

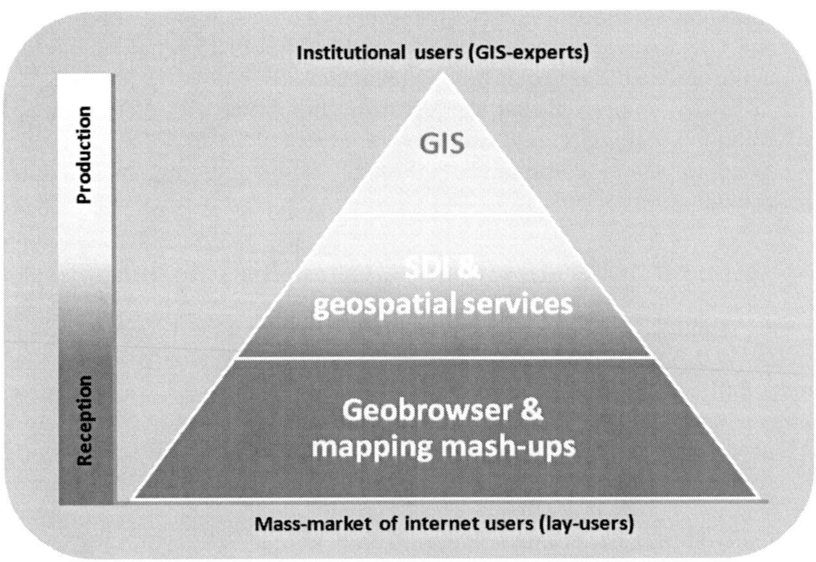

Figure 2-1: Geobrowsers serve the massmarket segment of Internet users, while GIS-experts produce and lay users consume geographic information (own illustration)

The shift of geomedia towards social software and geo-tagging lays the foundation for a 'democratisation of cartography' (Crampton, 2010, 37), i.e. the creation of a vast number of mapping mash-ups that bears upon everyday activities, and user-driven modes of production and co-production of geographic information. It is thus symptomatic of geomedia to absorb, or become conflated with, the social software domain, and let users share, collaborate and communicate through geographic information (Scharl & Tochtermann, 2007; Gordon, 2007; Crampton, 2010).

In order to organize these increasing streams of textual and non-textual information in online communities, and in web2.0 in general, geomedia-based communication became popular, allowing location to be the organizing principle (Lapenta, 2011). Various geomedia applications are linked to online communities, or are even entirely based on an approach to share, communicate and collaborate in online communities according to a common framework of orientation, i.e. geographically explicit references. These include text, audio, video or any other type of media that is linked to coordinates that reference a geographical territory by a geo-coded point, line or area. Geo-tagged communication in online communities radically changes 'the ways we visualize and understand the world around us – its places, geographies, and relationships' (Crampton, 2010, 37). Thereby it calls for an understanding of the practices of geo-tagging and how they are intertwined with everyday spatialities.

3 Mapping Becomes a Conversational Concept of Communicating

Geo-tagging for communication and cooperation in online communities is implemented on the level of the technological infrastructure by varying modes, degree and depth of prosumption of geographic information. It implies the negotiation of two key elements of spatial data that have, by then, been shaped by an institutionalized process: Location and Attributes (Schuurman, 2004) or 'properties z present at locations x', as Sui and Goodchild (2011, 1743) denote it.

Mapping as a process of communicating 'unary spatial knowledge' (Sui & Goodchild, 2011, 1743) from the expert to the user thus becomes a substantially more conversational way of communicating information between users. Those different, but not mutually exclusive, modes of prosumption extend the traditional broadcasting concept of producing geographic information driven by a linear map communication model (Crampton, 2010).

According to Schön et al. (2011), online communities can be divided into three classes dedicated to (i) sharing information, (ii) communication and (iii) collaboration. Applied to geomedia applications, these three modes of prosumption allow for a systematic overview of the conversational aspects of geomedia applications.

Sharing. Applications for sharing geo-tagged media resources are based on an approach similar to platforms for the sharing of photos (e.g. flickr.com), movies (e.g. youtube.com) or bookmarks (e.g. delicious.com). Active users share information about a location, such as their whereabouts, the situation on the roads on which they are driving, good supermarket offers, free parking places, emergency situations, and other observations about their everyday surroundings. All this information is captured and collected by the social location application and distributed to the online community in a process known as 'location-sharing' (Tsai et al., 2009). Facebook Places, Plazes.com, Dopplr.com and Google Latitude are exemple applications where users share their previous, actual or future whereabouts. Location-sharing is applied mainly in order to keep a geographical diary and to obtain information about the surrounding area, but it is also used to organize spare-time activities or meet up with nearby friends (Fischer, 2009).

Various social location applications exist for sharing outdoor activities in online communities, such as GPSies.com, Bikemap.net or Wandermap.net. In these online communities, active users share hiking and cycling routes, or jogging tours they have accomplished. All routes are compiled and made searchable by the platform operator. A multifaceted source of individual routes for outdoor activities emerges, enriching the structuring of the recreational experience (Fischer, 2011). An international stir was created by the application Ushahidi (ushahidi.com), a platform that informs 'crisis mapping' and empowers global citizen journalism by user-generated reports from political and areas of natural disasters, like Sri Lanka, Haiti, Japan or Kenya (Schenker, 2009; Bulkley, 2010; Naone, 2011). Crisis mapping is commonly understood as the creation, analysis and visualization of real-time data to support humanitarian response in crisis areas (Meier, 2009). Users of Ushahidi typically share reports from inside a crisis area by way of geo-tagged text messages and photos. This enhances a people-centred information collection for analysis and visualization by crisis reaction forces, help in the field, and an international audience. Generally the prosumptive activity of sharing establishes user-generated collections of geographic information that are published to a mainly unrestricted audience. However, the collections might be filtered according to the privacy issues of the contributing users, or they might be intensively processed by the platform operator before being published.

Communication. Communication as a form of prosumptive activity concentrates on the discussion, collection and exchange of information, opinions and experiences bounded by geographical reference. Hence it aims at communication and opinion-forming about distinct locations. Examples are local search applications such as Yelp.com, Qype.com, Google Places or CitySearch.com, where active users contribute comments and ratings about services and goods within the everyday living environment. Like local search applications, the recreational platforms Outdooractive.com, Tripadvisor.com and DeineBerge.de allow users to review and comment on outdoor activities and accommodation. Platforms for public participation regularly encourage user-centred communication about places for urban planning and urban management. For example, applications like Wildurb.at or YouXcity.com are implemented to facilitate public discussion and opinion-forming about distinct locations.

Collaboration. Applications can facilitate collaboration between users to create geographically referenced content or services in a concerted way. In doing so, active users pursue a common goal, intention or challenge. A notable example is the OpenStreetMap (OSM) project, founded in 2004. Contributors to OSM (openstreetmap.org) aim to create a map of the world at no cost that is usable for all. Members of the OSM community capture geo-data about anything of interest and compile it into a database. Another example is Wikimapia.org, a free editable map that allows users to add information for any location in the world. It combines Google's digital mapping with a wiki system. Users can mark out a polygon to tag locations, and link text, video or any other media to it. Another form of collaboration is introduced by location-based games. They incorporate geo-tagging to link up material space with game space. The players interact with each other through geo-tagged locations in order to fulfil a task or to challenge each other, as with the applications Tourality (tourality.com) or FastFootChallenge (fastfoot.mobi). This gamification of space reconfigures spatialities as it supports new arrangements of socio-spatial practice and changes existing spatial settings (Fischer, 2013).

Degree and Depth of Prosumption. There are three basic ways of prosuming geographic information: community-based, editorial-based and aggregation. In addition to platforms based solely on the contributions of online communities, some platforms combine prosumption with data from mapping agencies or editorial-based content, such as Falk.de, Bing, Twitter Maps (www.microsoft.com/web/solutions/bing-twitter.aspx), which aggregates and geo-references tweets from twitter.com, as Google Places does to a certain degree, from various prosumed information sources. Furthermore, there is a difference in the depth of prosuming geographic information: on the one hand the capturing of locations of people, objects and resources by means of geographic coordinates, and on the other, the addition and enrichment of attributes of locations by means of a description, classification or discussion. Platforms may focus on both or either (Fischer, 2010a). Plainly worded and with reference to the conceptional construct of a Point-of-Interest, either the point or the interest or both are put in the prosumers' main focus of activity. Both can then be subject to user-driven enrichment, discussion and negotiation by mode, degree and depth of prosumption.

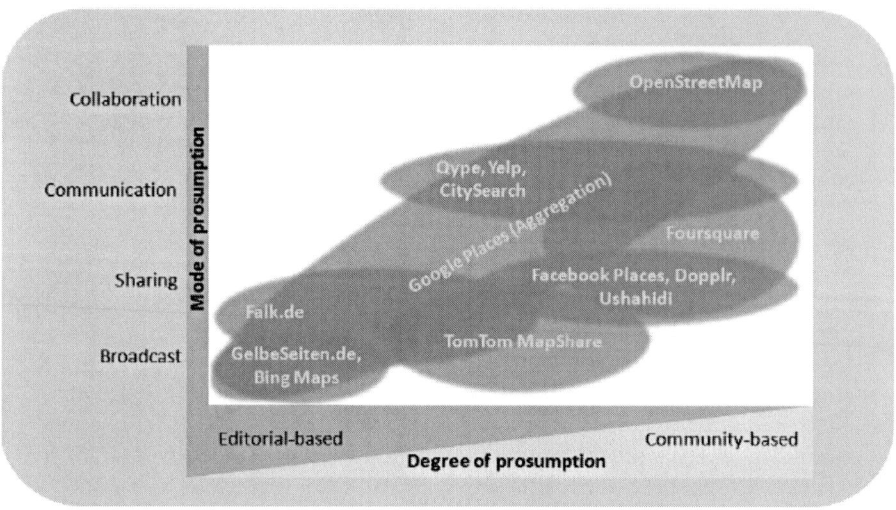

Figure 2-2: Classification of social location applications by mode and degree of prosumption (own illustration)

This overview depicts the range of models of cooperation and integration of user-generated geographic information in geomedia applications. It provides an understanding of how user-generated geographic information differs conceptually from, and may supplement, institutionalized processes of collecting geographic information. Furthermore, it elaborates how mode, degree and depth of prosumption provide a general framework for the participation and cooperation of users, and how these different forms of interactions are transformed into geographic information.

4 From Mapping Space to Coding Space

While these transformations concern the very nature of geoinformation for knowledge construction about space and its embeddedness in everyday contexts, geomedia facilitate an ever-increasing interweaving with everyday spaces: they are interfaces that drive interactions with any spatial resource. In this way geomedia applications expand the perspective of geovisualization from task-orientated information retrieval for spatial decision-making to infrastructures that make an urban space of networked places searchable, accessible, negotiable and perceptible, and thereby also co-constitute it.

Social location services are typified as a new aspect of this kind of application, grounded on participative practices of authoring, dissemination and consuming distinct geo-coded tags, narratives and discussions, as on yelp.com, foursquare.com or bikemap.net, for example. Thereby, social location services provide additional technologies to create new contexts of negotiation, visibility, memory and anticipation that change the socio-spatial arrangements in cities (Gordon, 2008). This assumption is underpinned by general research on geographical information systems (GIS) and geography (Harley, 1989; Wood, 1992; Pickles, 1995; Elwood, 2007). Based on their work, social location services can be seen as serving as an intermediary for urban activities that produce new codings, practices and possible geographies.

However, it is arguable how users adopt certain sets of functionalities in the context of everyday urban life, eventually enacting urban space.

Rob Kitchin and Martin Dodge (2011) developed a concept of *code/space* and *coded space* in order to understand the logic of new media and their interweaving with space. In a *code/space*, people depend on software to do things and cannot do without it. In the related concept of *coded space*, software modifies or extends the production and functioning of space. Without software the spaces continue to function as intended but without any additional features (Kitchin & Dodge, 2011). Urban space needs to be considered as a context, being (re-)created by the ongoing, everyday practices of its users, such as inhabitants, tourists, workers, students or shoppers (de Certeau, 1984; Lefebvre, 1991). Progressing from research on location-based services, urban space is considered by others as a context parameter of particular interest only (Reichenbacher, 2009). Paul Dourish (2004) describes this problematic positivist usage of the term 'context', stating that 'context cannot be a stable, external description of the setting in which activity arises' (Dourish, 2004, 2). According to Lefebvre (1991) and to Kitchin & Dogde (2011), space as a context is something people do and an outcome of the practice of everyday life.

As elaborated above, social location services provide different modes and degrees of coding space (and thereby of appropriating it) by user-generated geographic information. In the following paragraphs, this framework will be used to present the results of a study of the platform Qype.com in Berlin. The study examines practices of use, i.e. the ways of using the functionalities of Qype, and explores links between social location services and the lived spatiality of its users.

Editorial-based directory services create markets where indirect network effects between users and local businesses exist (Varian, Farrell & Shapiro, 2004). Their content is used by readers to orientate themselves in urban space, while local businesses pay the platform operators to be visible, as they expect readers to become customers. Hence editorial-based directory services are two-sided markets with two target groups: users and local businesses. The study will illustrate how a multi-sided market emerges with diverse links resulting from indirect effects and power relations among local businesses, Qype as a platform and users as customers, users as self-marketers and users who aim at social community forming. This transformation of a formerly two-sided market through the use of Qype induces new socio-spatial arrangements and diversifies the appropriation of urban space.

5 A Study of Qype.com

Qype provides a user-generated and geo-coded collection of Points-of-Interest (POI) that is searchable by address, district, topic and category of location. Registered users can add new locations, rate and review locations, and socialize through messages and forum discussions. In order to reconstruct how Qype is woven into the users' lived space of Berlin, 19 topical, in-depth interviews (Kromrey, 2006) were conducted with selected users of Qype over a period of four months. The interviews covered the topics of everyday urban life in Berlin, as well as uses of Qype and general usage of media. In the course of the study, recurrent themes were identified and categories were generated inductively. Using the constant comparative method (Strauss & Corbin, 1999), analysis of the interviews revealed five main themes that concern the appropriation of Qype in Berlin. The first two are urban contexts of using Qype that were condensed into the concept of 'localizing capital' during the subsequent interpretation. They shape three main practices directed towards mastering urban contingency, social community forming, and self-marketing which enact urban space.

Localizing Capital. The quality of a place as a means of identity formation turned out to be central for the use of Qype. Cafés, restaurants, clubs, bars and shops provide unambiguous, symbolic worlds for social integration and identity formation. In these settings, diverse lifestyles condense and enable a sustainable stabilization and expression of identity (Bertels, 1990; Weichhart, 2008). In this respect, the users depicted what Bell and Valentine (1997) denote as 'we are where we eat' when talking about the neighbourhoods in Berlin. But the users acknowledge that local businesses as producers of symbolic values hold different affordances. They might be divided into ego-signs that provide great opportunities for individual differentiation, or, rather, provide a universally accepted utility function, like a bakery. The latter are less likely to be searched and reviewed on Qype.

Furthermore, the interviews reflected that the interviewees' respective quotidian scope of action is shaped by different levels of fragmentation, flexibility and dissolution. Commuting activities and fragmented everyday spatial practice are commonly accompanied by a frequent experience of urban spaces that are geared to pure functionality and bare use, for example airports, train stations and footbridges (Augé, 2010). Interviewees who are often localized in these spaces in order to connect home, work and leisure feel restricted in their access to meaningful places of consumption.

The resulting individual capacity of self-determined localization was termed 'localizing capital', which refers to a perspective on habitual urbanity, a theory of urbanity based on Pierre Bourdieu's social theory (Dirksmeier, 2009). Referring to Bourdieu's (1984) theory of practice, 'habitus' is a result of learning processes and facilitates social actors' production of social practice. It unfolds within the boundaries of the available share of economic, social, cultural and symbolic capital to cope with the contingency of urban life. The competition for a share of capitals in a field produces discernible differences in appropriated space. Different shares of capital are reified and naturalized in space (Dirksmeier, 2009).

In the case of Qype, this field is urban consumption and gastronomy in particular. It is a field of local business owners, intermediaries and users as consumers who are connected by mutual relationships concerning the exchange of monetary, cultural and symbolic capitals. Localizing capital then emerges from the mutual interdependence of place-specific knowledge as 'informational capital' (Bourdieu & Wacquant, 1996) and the users' quotidian spatio-temporal scope of action.

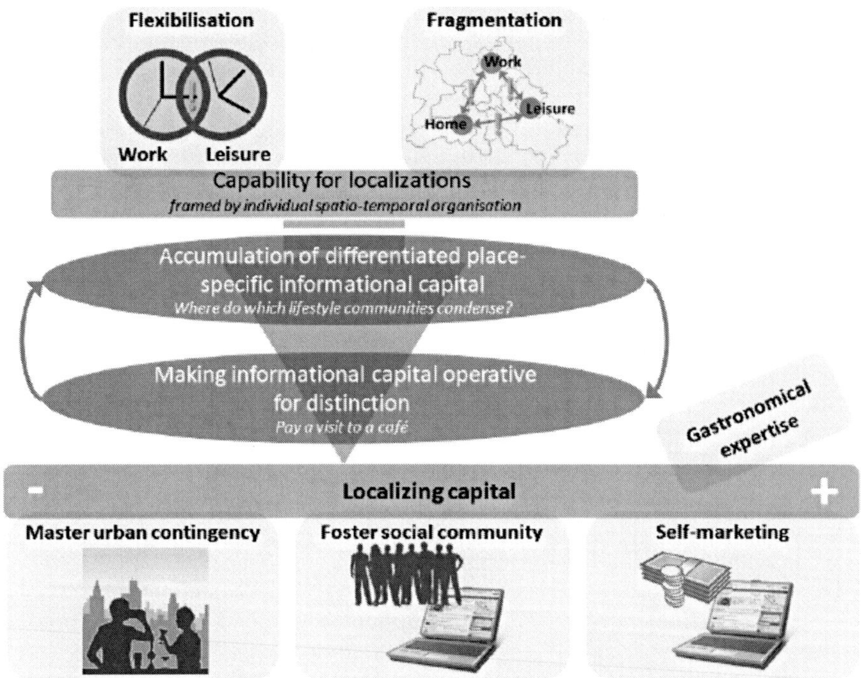

Figure 2-3: Accumulation and handling of localizing capital on Qype

Localizing capital is acquired individually in a process of learning that requires time and labour (Bourdieu, 1984), for example by visits to various cafés and the differential experience of consumer practices, goods and their symbolic values on-site (see Fig. 2-3 – middle section). However, to become effective, informational capital must be recognizable for others by localization. Hence, localizations in urban space are also a means of acquiring informational capital and make informational capital operative. Furthermore, the capability for localizations is framed by individual spatio-temporal fragmentation, flexibility and dissolution, forming the localizing capital (see Fig. 2-3 – upper section).

Hence the approach of habitual urbanity links the use of Qype to the 'symbolic economy' (Zukin, 1995, 8) of Berlin by the concept of localizing capital. A different level of localizing capital and a different way of processing it (see Fig. 2-3 – lower section) result in three main types of use-practices (i.e. the ways of using a set of functionalities of Qype). They are directed towards mastering urban contingency, social community forming, and professional self-marketing. Most users of Qype are not representative of one distinct practice but rather combine two, or even all of them. Hence the types of practices are considered domains on a continuum that is inhabited by users to varying extents.

Mastering Urban Contingency. Users who hold minor localizing capital to find places that provide prospects for individual distinction make use of ratings, reviews and the profile information of the respective reviewers. Regarding lifestyle, consistently positive ratings are considered homogeneous places whereas strongly divergent ratings indicate conflicting places. The users consider the style of writing and the core issues of the reviews, and attempt to assess the habitual character of a reviewer and her cultural competence to contribute:

> I think it's pretty interesting to see the kind of impression you can get about people just by reading their reviews and comments. How you can get an idea of how they are. […] Up to a certain point, you can see what kind of people they are just by reading their comments.

Qype catalyses the users' minor localizing capital as it bypasses efforts for direct discovery and trial-and-error of local businesses in urban space. Instead, the users adapt objectified informational capital from reviews to optimize tangible appropriations of urban space.

Social Community Forming. For numerous users, the online community of Qype is a means of high relevance for 'socializing' (Magid, 2007). Their time for leisure activities is filled with visits and review activities. That is, they are able to maximize localizing capital within their everyday routines. Visiting cafés or restaurants becomes a passion and they write reviews about nearly every place they visit. Thereby they objectify informational capital by authoring their experiences on Qype:

> Regardless of where I've been, I always do a review of it afterwards. I focus first and foremost on writing my own reviews.

A passion for writing is typical. However, the users normally lack expertise in gastronomy and professional literacy. They consider obtaining compliments, points and badges for reviews as important for eventually being recognized by users who share their passion and socializing through Qype's online community. The objectified informational capital is converted into institutionalized capital by Qype's reward functionalities: compliments given by other users, and the status-points and badges given by the platform operator for authoring activities. In this way, Qype institutionalizes that passion and creates and fosters new formations of social communities through locative performances. As a result, the relationship to local businesses as producers of resources for spatially related identity formation changes fundamentally. Users of this kind, rather, conceive places of consumption as collectibles that serve individual social community building, by collecting a first review of a local business, a review of a classic location, or one of a place also reviewed by friends.

Professional Self-Marketing. Becoming a professional 'Qyper' requires an intrinsic means of production and a high amount of localizing capital that might not be available to everyone. It complements the professional occupation of users, sometimes becoming a professional occupation in itself. The users invest immense localizing capital as a means of production, like a professional skill in writing, or a professional expertise in gastronomy, to make reviews of specific, local businesses a part of their portfolio by using Qype:

> I've got my own project. I write at least once a month whenever we go restaurant hopping. The review goes to the open press and then onto Qype. That's the minimum amount that I write.

This practice is orientated towards self-marketing and the consolidation of the user's professional identity as a personal brand. As soon as their symbolic capital becomes recognized by local businesses, users can convert it into economic capital by means of a new business relationship or even direct benefits, e.g. a financial or in-kind reward for a review (see Figure 2-4: dotted line at the bottom).

In contrast to editorial-based directory services, these three practices of Qype users are intertwined by interactions on the level of local businesses and Qype's platform. Users sort reviews to catalyse low levels of individual localizing capital and facilitate territorial relationships (Figure 2-4: middle section).

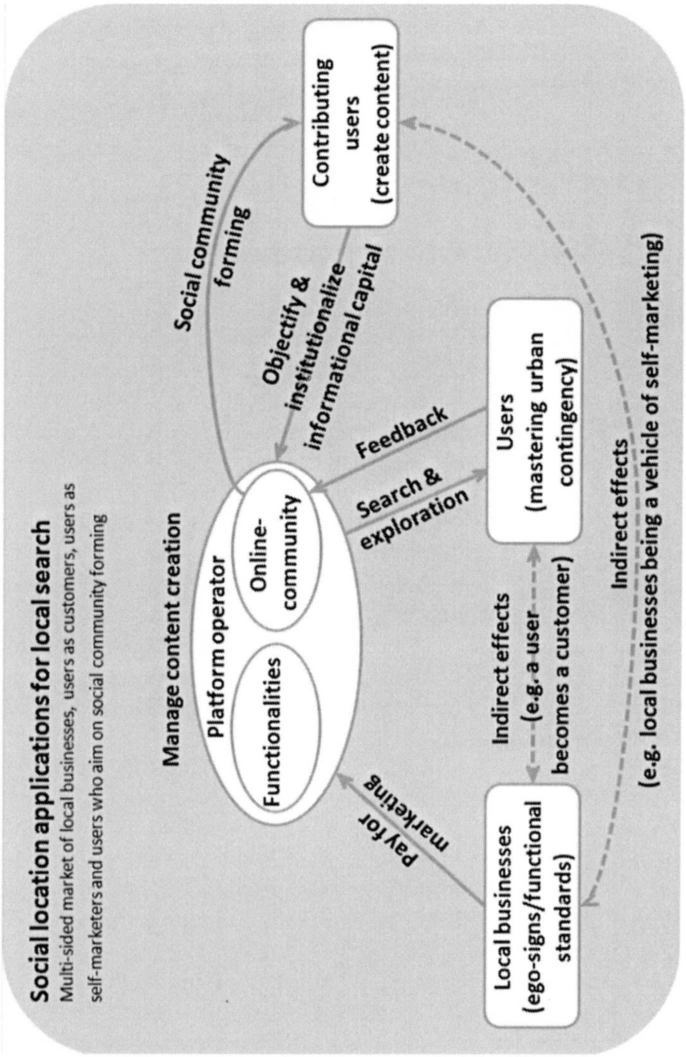

Figure 2-4: Relations between users, local businesses and Qype in a multi-sided market

The review functionality is used by passionate and professional contributors to objectify their high levels of localizing capital (see Figure 2-4: right-hand side). In turn, reward functionalities, such as status-points and compliments given for reviews, reinforce the symbolic capital of contributing users, who convert it into economic and social capital. These functionalities connect users with low and high levels of localizing capital. It allows a 'community of practice' (Lave & Wenger, 1991) to emerge, with users mastering urban contingency on the margin and professionals who valorize their localizing capital in the centre.

6 New Socio-Spatial Arrangements

The use of Qype, particularly the adopted set of functionalities, can introduce new ways of creating relationships between people and places, other than visiting a place, and new forms of relationships between people and places, other than being a customer only. Qype thus drives the creation of new socio-spatial arrangements that transform the lived spatiality of users. Three main transformational aspects emerged from the research process.

First, the use of Qype goes far beyond filtering by physical proximity alone. The users of Qype attempt to evaluate and internalize the symbolic meaning attached to locations by profile, status and attitude of users who contributed reviews. These attributes of places support urban activities in the differentiated and overlapping territorial spaces of lifestyle communities in a city. In brief: Qype is an interface that makes a networked and differentiated urban heterogeneity visible and accessible for users with low levels of localizing capital.

Secondly, the relevance of distance-driven locational factors might be redeployed as physical proximity fades into the background during the exploration of public space. Distance-driven locational factors of local businesses might therefore become less relevant for the user's choice. Hence the relevance of centrality, reachability and visibility in public space might be softened in favour of findability, the attraction of different reviewers, and check-ins on Qype.

Thirdly, contributing users gain substantial power and mediate between local businesses and customers. Hence, localizing capital and the applied functionalities for its conversion shape a new digital economy of place: contributing users and platform operators authoritatively govern the interface and negotiate new meanings of space. User-disempowerment in public space might happen to those who are not able to contribute because of a lack of means of production or of access to the technology. Hence Qype might induce new issues of exclusion and control of public space.

7 Considerations about Everyday Geomedia Use and the Appropriation of Space in the Educational Context

The concept of localizing capital supports an integrated view of spatial contexts in which users and social location services are embedded. It can be applied to consider the use of certain sets of functionalities in the urban context as a socio-spatial process. It links the use of functionalities to processes of catalysis and conversion of capital in a symbolic economy of a city, and therefore connects to the lived spatiality of a city. Hence, localizing capital provides a basis for thinking about the individual actions of users of social location services that allow the emergence of collective socio-spatial arrangements; it can also be used to underpin thinking with regards to business cases and governance strategies for socio-spatial arrangements in the community management of social location services.

It might be difficult to generalize from this particular case study on Qype to everyday geomedia use. Nevertheless, it may be assumed that similar processes of utilization and valorization of localizing capital exist in other fields of action. These will be a matter of the actual mode and degree of prosumption used to address aspects of the appropriation of space by geomedia. An example might be popular portals for outdoor activities, like Bikemap.net or GPSies.com, where users collect interesting routes for cyclists, joggers and hikers. Likewise streets and cycle ways can be symbolically appropriated and consumed by cyclists and joggers. An imagined two-sided market, then, consists of the cyclists and the providers of cycle ways (in most cases the government). However, the provider does not earn money from the consumption of cycle ways, but there may be other benefits (e.g. to making cycling more attractive and thus changing the modal split). A direct valorization for an external platform operator appears less likely, however. Presumably, the respective platform operators implement advertising-based business models rather than marketing-based business models on that account, although passionate and even professional contributing users might exist and benefit. Professionals might, for example, use the symbolic capital from their activities on the platform to underline their professional competence as tour guides or writers on outdoor activities, for their self-marketing in a professional occupation.

Thereby geomedia applications might become a supplement for aspects of tangible urban governance, a form of cooperative management of urban public space, empowering users and activating their resources (Frey, 2009; Benz & Dose, 2010). Hence it becomes important for education to mediate these new specific dealings with urban space by geomedia, and to enable an emancipatory and responsible appropriation of space. The case study illustrates that we need to push the subject of technology to the core of the curriculum and recognize the practices, the inclusion and the enhancement of informational capital and new spatial relationships that can be strengthened by geomedia. Educators need to consider these utilizations of geomedia in order to inform students about, and empower them with, these new modes of our linkage to spatial meanings. With this geospatial perspective in mind, education can support an emancipatory and critical everyday use of geographic information in a networked society, and mediate methods and skills to empower citizens' possibilities to participate responsibly in society.

References

Augé, M. (2010): *Nicht-Orte*. München: C. H. Beck.

Bell, D. & Gill, V. (eds.) (1997): *Consuming Geographies: We Are Where we Eat.* London: Routledge.

Bertels, L. & Herlyn, U. (eds.) (1990): *Lebenslauf und Raumerfahrung*. Opladen: Leske-Budrich.

Bourdieu, P. (1984): *Die feinen Unterschiede: Kritik der gesellschaftlichen Urteilskraft*. Frankfurt am Main: Suhrkamp.

Bourdieu, P. & Wacquant, L. J. D. (1996): *Reflexive Anthropologie*. Frankfurt am Main: Suhrkamp.

Bulkley, K. (2010): 'Mobile Technology Takes Centre Stage in Disaster Relief'. *The Guardian* (June 18 2010). Retrieved from: http://www.guardian.co.uk/activate/mobile-technology-disaster-relief.

Carl, D., Clausen, J., Hassler, M. & A. Zund (2008): *Mashups programmieren.* Köln: O'Reilly. Retrieved from: http://books.google.at/books?id=GpHBV4zNDUsC.

Crampton, J. (2010): *Mapping: A Critical Introduction to Cartography and GIS.* Chichester: Wiley-Blackwell.

de Certeau, M. (1984): *The Practice of Everyday Life*. Berkeley: University of California Press.

Dirksmeier, P. (2009): *Urbanität als Habitus: zur Sozialgeographie städtischen Lebens auf dem Land*. Bielefeld: transcript.

Dourish, P. (2004). 'What we Talk About when we Talk About Context'. *Personal and Ubiquitous Computing* 8 (1), 19–30

Elwood, S. (2007): *Critical GIS Perspectives on Volunteered Geographic Information.* Washington DC. http://www.ncgia.ucsb.edu/projects/vgi/docs/present/Elwood_VGI_2007.pdf.

Fischer, F. (2009): 'Learning in Geocommunities: An Explorative View on Geo-Social Network Communities'. In T. Jekel, A. Koller & K. Donert (eds.), *Lernen mit Geoinformation IV*. Heidelberg: Wichmann, 12–21.

—. (2010a): 'Wertschöpfung 2.0: Neue Produktions- und Nutzungspraktiken auf dem Geoinformationsmarkt'. *GW-Unterricht* 120, 30–46.

—. (2010b): 'Everything will be Geo-Tagged. Local Search Media Become Social and Mobile in 2010 Finally'. *Geoinformatics* 1 (13), 10–12.

—. (2011): 'Tourism Mapping. Digital Geomedia vs. Paper Maps'. *Geoinformatics* 1 (14), 10–12.

—. (2013): 'Playful Augmentations of Space: Differenzerfahrungen im urbanen Raum?'. In I. Gryl, Nehrdich T. & Vogler R. (eds.), *Geo@web: Medium, Räumlichkeit und Geographische Bildung*. Berlin: Springer, 127–144.

Gordon, E. (2007): 'Mapping Digital Networks. From Cyberspace to Google'. *Information, Communication & Society* 10 (6), 885–901.

—. (2008): 'Towards a Theory of Network Locality'. *First Monday* 13 (10), http://firstmonday.org/ojs/index.php/fm/article/view/2157/2035.

Harley, J. B. (1989): 'Deconstructing the Map'. *Cartographica* 26, 1–20.

Kitchin, R. & Dodge, M. (2011): *Code Space: Software and Everyday Life*. Cambridge, MA: MIT Press.

Kromrey, H. (2006): *Empirische Sozialforschung: Modelle und Methoden der standardisierten Datenerhebung und Datenauswertung*. (11th edition). Stuttgart: Lucius & Lucius.

Lapenta, F. (2011): 'Geomedia: On Location-Based Media, the Changing Status of Collective Image Production and the Emergence of Social Navigation Systems'. *Visual Studies* 26 (1), 14–24.

Lefebvre, H. (1991): *The Production of Space*. Oxford: Blackwell.

Magid, L. (2007): *MySpace Unraveled : A Parent's Guide to Teen Social Networking from the Directors of BlogSafety.com*. Berkeley, CA.: Peachpit Press.

Meier, P. (2009): 'Proposing the Field of Crisis Mapping'. *iRevolution.* http://irevolution.net/2009/08/08/proposing-crisis-mapping/.

Naone, E. (2011): 'Internet Activists Mobilize for Japan'. *Technology Review*, http://www.technologyreview.com/communications/35097/?p1=A1&a=f.

Pickles, J. (1995): *Ground Truth: The Social Implications of Geographic Information Systems*. New York: Guilford Press.

Purvis, M. & Svennerberg, G. (2010): *Beginning Google Maps API 3*. New York: Apress.

http://books.google.at/books?id=gHb8m0GSV2MC.
Reichenbacher, T. (2009): 'Geographic Relevance in Mobile Services'. In E. Wilde, S. Boll, K. Cheverst, P. Fröhlich, R. Purves & J. Schöning (eds.), *Proceedings of the 2nd International Workshop on Location and the Web – LOCWEB '09*. Boston, MA: ACM. http://portal.acm.org/citation.cfm?doid=1507136.1507146.
Scharl, A. & Tochtermann, K. (2007): *The Geospatial Web: How Geobrowsers, Social Software and the Web 2.0 are Shaping the Network Society*. Heidelberg: Springer.
Schenker, J. (2009): 'Ushahidi Empowers Global Citizen Journalists. *Bloomberg Businessweek*. http://www.businessweek.com/stories/2009-12–07/ushahidi-empowers-global-citizen-journalistsbusinessweek-business-news-stock-market-and-financial-advice.
Schön, S., Wieden-Bischof, D. & Schneider, C. (2011): 'Mobile Lerngemeinschaften'. In S. Schön, D. Wieden-Bischof, C. Schneider & M. Schumann (eds.), *Mobile Gemeinschaften: erfolgreiche Beispiele aus den Bereichen Spielen, Lernen und Gesundheit*. Salzburg: Salzburg Research, 61–80.
Schuurman, N. (2004): *GIS: A Short Introduction*. Malden, MA: Blackwell. Pub. http://books.google.com/books?id=r8Lzw_GgzS0C.
Strauss, A. L. & Corbin, J. M. (1999): *Grounded Theory: Grundlagen qualitativer Sozialforschung*. Weinheim: Psychologie Verlags Union.
Sui, D. & Goodchild, M. (2011): 'The Convergence of GIS and Social Media: Challenges for GIScience'. *International Journal of Geographical Information Science* 25 (11), 1–12.
Tsai, J. Y., Kelley, P. G., Cranor, L. F. & Sadeh, N. (2009): 'Location-Sharing Technologies: Privacy Risks and Controls'. *A Journal of Law and Policy for the Information Society* 6 (2), Summer 2010, 119–151.
Varian, H., Farrell, J. & Shapiro, C. (2004): *The Economics of Information Technology: An Introduction*. Cambridge & New York: Cambridge University Press.
Weichhart, P. (2008): *Entwicklungslinien der Sozialgeographie. Von Hans Bobek bis Benno Werlen*. Stuttgart: Franz Steiner Verlag.
Wood, D. (1992): *The Power of Maps*. New York: Guildford Press.
Zukin, S. (1995): *The Cultures of Cities*. Cambridge, MA: Blackwell.

CHAPTER THREE

EDUCATIONAL USES OF GEOMEDIA

INGA GRYL, ERIC SANCHEZ, THOMAS JEKEL, CAROLINE JOUNEAU-SION, JOHN LYON AND STEFFEN HÖHNLE

1 Challenges and Opportunities

In 1928, the *New York Times* published a picture of a geography lesson being taught in an aeroplane. The depiction of such an 'aerial geography lesson' underlined the fact that the most promising and powerful technologies available could foster the learning process and easily enrich and improve everyday life, particularly in educational and future-orientated contexts. Indeed, the plane offered the opportunity to change the point of view in a similar way that digital globes do today. Unfortunately, a brief look at the picture illustrates that it is doubtful that the course succeeded in exploiting the added value of the technology. It is apparent that the students do not seem to be looking through the windows (with the exception of the girl in the back of the classroom) and the teacher is showing a globe rather than the landscape below. In addition, due to the noise of the engines, the situation seems unlikely to have been particuluarly comfortable.

Today, many teachers face a technologically different, yet similar, situation. On the one hand they have the opportunity to benefit from powerful geomedia technologies to improve their lessons, but on the other hand they have to deal with problems and difficulties handling and making sense of them. Nevertheless, remaining in ignorance is not a realistic option at all as, in contrast to the example listed above, today geomedia technologies pervade everyday life much more than the aeroplane did in the 1920s. Challenges and opportunities emerging with the societal impact of geomedia need to be addressed through education. Indeed, the omnipresence of geomedia, their accessibility and power to shape action and perception, and the economic value of geo-data clearly require the implementation of geomedia utilization. Geomedia certainly do not change the learning processes themselves without careful implementation, and neither is the mature valorization of geomedia as tools useful in everyday social life an automatic assumption.

New York Times No. 306-NT-520A-6 in National Archives

"To-day's Aerial Geography Lesson"

Figure 3-1: An 'aerial geography lesson' in 1928, article from the *New York Times* (National Archives).

Therefore, this book presents a range of educational approaches and ideas around learning with and about geomedia, taking into account the potential as well as the risks and challenges associated with the use geomedia in education. This includes understanding the inherently biased nature of maps and other spatial representations; the power of geomedia produced by institutions, as well as the power of geomedia produced by lay persons; the use of geoinformation in the control and surveillance of individuals and groups, and the opportunities of georeferenced data to foster innovative knowledge. Addressing these aspects in essence means preparing students for an emerging Geoinformation society. During recent decades, the focus, aims and conceptions of this preparation have changed significantly.

2 Geoinformation Use in Education: A brief Roundup of Approaches

Interest surrounding the introduction of GIS in education can be traced back to the mid 1990s (Milson, Kerski & Demirci, 2012). This interest was driven by enthusiasts from technology and industry backgrounds, as well as by a set of innovative and technology-orientated teachers, who embraced the possibility of exploring 'real world mapping' and 'real world research' with their students. This work had hitherto been the realm of higher education, research institutions and government authorities. Several of the early enthusiasts also suggested that the use of digital geoinformation (GI) would enhance awareness of environmental problems and therefore promote responsible action with regard to environmental concerns. These assumptions waned in the light of practical considerations as the transfer of GI-Systems to schools was seemingly not viable or achievable beyond a set of very innovative and confident teachers. It has also been pointed out that awareness and action are not directly linked (e.g. Johannson, 2006; Kerski, 2003; Milson, Kerski & Demirci, 2012; Baker et al., 2012). However, it can be said from an analysis of European curricula that this rather 'technology focused' approach has succeeded in implementing GIS at various levels of education, and has given 'spatial subjects' (e.g. geography and environmental education) a more modern link to the world beyond education in general and school administrators in particular.

On a more systematic note, we may identify four recent arguments for the inclusion of GI in secondary education. These arguments also reflect the development from GIS-use in school to geomedia use in school, and therefore to an increasing orientation for everyday uses. This development is very much in line with both the technical and societal developments outlined by Josef Strobl and Florian Fischer in this volume.

These arguments are (a) a technical argument, e.g. that GIS might be included in school curricula to prepare students for the requirements of the increasing geospatial workforce; (b) the argument of spatial thinking as a key competence for problem solving in a variety of subjects; (c) an argument that refocuses on GI as a subject in secondary education and on its everyday use (Gryl & Jekel, 2012), and (d) the argument of spatially enabled learning (Vogler et al., 2012).

Enlarging the Geospatial Workforce. The technical argument for the inclusion of GI in secondary and undergraduate education was developed mainly in the US, but has since been followed across Europe. It roughly argues that (secondary) education should be geared towards workforce needs, preparing students to join this workforce (see Kerski, 2003; Donert, 2010). Following this basic idea, educational models were developed that place strong emphasis on industry-related skills (Body of Knowledge: DiBiase et al., 2006; 2010, Schulze, Kanwischer & Reudenbach, 2011).

This educational model was largely translated to secondary education in the early stages of GI inclusion in school, providing spatial subjects with a new *raison d'être*. However, the approach can be questioned on several accounts. First of all, criticism can be made of teaching lower-level GIS competencies and imparting often short-lived software skills; in addition, the commercial objectives of the private sector cannot be a key driver in a secondary education which aims at providing students with competences needed for social inclusion and citizenship. Secondly, the development and simplification of software together with the emergence of digital globes have enabled learners to devote more time to basic concepts and critical thinking (Hakley, 2010, 87; Steinmann, 2013), rather than looking into industry-strength software systems. Additionally, although the technical design of geomedia in the future is largely unknown, it is to be expected that usability will more likely increase than decrease. Consequently, in this volume, the editors and authors clearly advocate a broad definition of (general) education with geomedia. Minor technical skills, such as the ability to use specific, appropriate software, are regarded as a necessary objective, but not as the aim of geomedia education.

Learning to Think Spatially. 'Learning to think Spatially' (National Research Council, 2006) argued for a strong inclusion of GIS in curricula on the assumption that spatial thinking (in terms of absolute space) would enhance the quality of decision-making in society and particularly in the sciences.

Thinking spatially (National Research Council, 2006) is a wider concept than the technological argument outlined above. It draws on three aspects, namely (a) concepts of space, (b) tools of representation, and (c) processes of reasoning. Together, it is argued, these dimensions help in solving a variety of spatial problems and should therefore be taught at secondary school level. Empirical research has led to a host of studies (Spatial Thinking Lab: Hegarty, 2011; Marsh, Golledge & Battersby,

2007; Tu Huynh, 2008), resulting in recommendations for better learning environments.

The concept of spatial thinking can be characterized from various perspectives. In sum, spatial thinking considers a comparatively narrow concept of space, namely the conception of absolute space and a quantitative spatial approach. Therefore, it can reasonably be part of science education in the field of the physical world, concerning particularly hypothesis generation through the spatial representation of phenomena. In this respect, it has its undisputed strengths and should be taught at secondary school level.

However, general education needs to include the social realm – the realm of learning and politics, conflicts, discourse and identity. Therefore, relational conceptions of space – explicitly excluded in the NRC report – are essential (Jekel, Pernkopf & Hölbling, 2008) in the humanities. We argue that Geoinformation *should* be included in the humanities, and especially in citizenship education, as a core dimension of digital humanities (Douhei, 2011). However, with the current conception of absolute space used in GIS in education, it may be argued that GI inclusion is simply seen as utilitarian, preparing students for a job market in the sciences, as the form of spatial reasoning proposed bears little resemblance to the arguments around space within the humanities (see Schuurman, 2004).

In addition, empirical research (Jekel et al., 2008) has shown that no links could be identified between learners' absolute conceptions of space on the one hand and their creativity and problem-solving capability in the political realm on the other (e.g. spatial planning). This calls for a wider conception of GI use in schools, including multiple spatialities beyond absolute space.

Spatial Citizenship. Spatial citizenship (Gryl, Jekel & Donert, 2010; Gryl & Jekel, 2012) re-centres the inclusion of geomedia on the everyday lives of individuals, rather than on the industrial workforce or science requirements. The concept looks into the everyday use and effects of spatialities, with specific reference to the now ubiquitous GI applications, as described in the contributions of Fischer and Strobl. The aim is to develop a reflexive practice with regard to spatial information and representation at the level of the individual. The concept therefore includes traditional competences of map reading and information access that have recently significantly changed due to technological development. Additionally, education for spatial citizenship aims at enabling secondary students to devise alternative spatial scenarios and to participate competitively in society with the help of geomedia.

The concept of Spatial Citizenship is based on three main theoretical traditions that enlarge traditional concepts of citizenship:
(1) the social appropriation of space as basis for participation in society,
(2) a critical perspective on spatial information and representations as discussed in the critical cartography / critical GI Science debate,
(3) various conceptions of citizenship education with special emphasis on emancipated approaches. Spatial Citizenship requires a consistent use of constructivist concepts in learning environments related to conceptions of space and the discipline of the subject itself.

Towards Spatially-Enabled Learning Across Subjects. Just as everyone today uses office programs for writing and calculations, we suggest that in the near future a spatial background will be part of many learning environments across subjects, even beyond the explicitly spatial domain. A spatial reference may provide an additional frame of reference and contextualization, even for topics that are not necessarily spatial in their outline. For example, novels are often more or less explicitly spatially referenced, being set in specific places and times, allowing geomedia to support literature studies. The same can be said of the arts, history and other subjects, which may use the spatial context as an additional frame of reference for learning (Vogler et al., 2012). The approach is clearly linked to ideas of dual coding (Paivio, 1982), using space as an additional form of coding in the learning process. This development will be supported hugely by ubiquitous mobile devices. Co-operations in this regard have been demonstrated by Schauppenlehner et al. (2013), who created virtual land art in Vienna using GPS drawings; Gruber (2007) provided an example for spatializing a novel by Christoph Ransmayr in an attempt to enhance students' understanding of modern Austrian Literature, enriching the reading experience by adding both recent and historical data on digital globes. In an interdisciplinary student project, Demirci and Gryl (2012) showed how the mapping of former homes of Holocaust victims illustrates the horrors of the genocide in the heart of society.

This approach is central to the notion of a spatial dimension and thinking in spatialities as essential contributions to inspire and enrich learning in various fields, for humans are geo-localized beings. While spatially-enabled learning currently remains a vision for the near future, several of the learning environments presented here may also be considered basic tools for this approach.

To conclude, this book discusses approaches to learning geared to promoting 'thinking spatially', spatial citizenship and spatially-enabled learning, while maintaining basic technical competences and paying special attention to the social realm of geomedia.

3 Pedagogies and Geomedia

Beyond the media-specific approaches described above, the general impacts of geomedia on learning and particularly their values for educational purposes can be outlined.

The pedagogical value of geomedia was identified early on by educators. For example, in 2007, a French study (Sanchez, 2009) of 862 geography and geology teachers emphasized the high interest expressed by teachers in the use of geomedia. The main focus was on the use of virtual globes.

A quick overview of the use of geomedia in education demonstrates that geomedia have been integrated within a wide range of pedagogical settings. This diversity relates to different pedagogical approaches, as detailed below.

Model-Based Learning (MBL) is grounded on the idea that scientific models are the core of scientific investigation (Halloun, 2004). As a 'simplified representation of a system, which concentrates attention on specific aspects of the system' (Ingham & Gilbert, 1991), a scientific model allows for

(1) the visualization and understanding of complex and hidden phenomena,

(2) communicating ideas and understanding of the lifeworld,

(3) making predictions by using simulation.

The desire to give models and modelling an important role in science education (Clement, 2000) demonstrates an increasing recognition of the role of models during the different reforms of scientific instruction in many countries. MBL is well illustrated by the work of Ludovic Delorme, a French secondary teacher. He shows how his students collected data about a watershed in the South of France both with the help of Google Earth and during a field trip. Afterwards, they created a digital model of the watershed with simple mapping tools and used this model to predict and discuss flood hazards and how land use management could take this risk into account. MBL is often linked to mapping spaces and opens up the possibility of evaluating students' mapping activities.

Problem-Based Learning (PBL) is a pedagogical approach based on the philosophy of Gaston Bachelard: knowledge is a repository of questions, and without questioning scientific knowledge would not exist (Bachelard, 1938). The epistemology of Bachelard has had an important impact on student-centred teaching practices, focusing on the commitment of students in solving more or less realistic problems. The pedagogical website of the Académie de Montpellier, France (http://svt.ac-montpellier .fr/spip/spip.php?rubrique10) dedicated to the use of geomedia in secondary education provides numerous examples: students may work on the consequences of the construction of wind turbines as well as the impact of over-fishing on the Bluefin tuna population. In these examples, geomedia enable students to address problems by using local or global geo-data.

Inquiry-Based Learning (IBL) is close to PBL. IBL is defined as an 'approach to learning that involves a process of exploration which leads to asking questions and making discoveries in the search for new understandings' (National Science Foundation, 2000). The role of IBL has been emphasized by a recent report from the European Commission (Rocard et al., 2007). Geomedia can provide the data and tools needed for such an inquiry. As an example, Van der Schee and Horvath (2006) describe a lesson designed for secondary school pupils who used GIS to investigate the location of coffee shops in their city, The Hague. The students mapped out schools and coffee shops by using public data, produced maps which were then combined, and drew buffers around the coffee shops. One interesting result of this study carried out by the students was the discovery that the law which states that a coffee shop should not be located within 500 metres of a school is infringed by numerous coffee shops, which then leads to various questions about market regulation.

Project-Based Learning is grounded on the philosophy of John Dewey (1938), which emphasized that learning is closely linked to doing. It encompasses both PBL and IBL, with a further focus on self-directed learning. Students are required to solve a problem, by carrying out an inquiry. In this specific pedagogical approach, the use of geomedia allows the learner to address non-deterministic, authentic and complex problems, close to reality, at a local or global level. Students collaboratively use the properties of space as a vehicle for asking questions, structuring problems, finding answers, and expressing solutions to 21st-century issues, such as global warming, globalization, land use management and everyday praxis.

The principles of PBL lie in the authenticity of the project. Sébastien Cathala, a geography teacher, reported on several projects that he carried out with upper secondary school students. One of these relates to the waste management for a school. The students were required to create a digital map of the dustbins of the school in order to propose new additional sites. The board of the school decided to implement the solution designed by the students in order to solve the problem of waste lying on the floor.

Game-Based Learning is similar to Project-Based Learning. Geomedia enables the design of realistic learning situations but, in this specific pedagogical approach, the consequences of the decisions are minimized. While the player takes decisions and the results of the game depend on these decisions, mistakes and failures are generally permitted and it is always possible to try again. With this type of scenario, uncertainty remains about the eventual outcome, which is open and not foreseen or pre-scripted by the teacher. The freedom to take decisions offered to the player is limited only by the rules that are the norms and conventions shared by the different participants. Sanchez and Jouneau-Sion (2009) describe a simulation designed with geomedia, where the situation relates to the design of a high speed railway loop dedicated to testing high speed trains (TGV). 20 students from a lower secondary school play the roles of different characters, such as citizens living close to the project who do not want their environment being partly destroyed, citizens looking for a job, a reporter for a local TV channel, local political authorities, a director of a company, and others. The game or simulation consists in finding a compromise for the successful implementation of the project. The students utilize a set of data in a virtual globe (for example Google Earth) to design a convincing set of arguments. PBL and GBL both promote the importance of developing citizenship as a participatory process.

4 Outlook and Invitation

The diversity of opportunities enabled by the use of geomedia is not limited to the previous examples. Further new perspectives in education are fostered by the use of geomedia based on web2.0 technologies. In sum, three main educational trends driven by technological and social changes have been identified in a recent report from the European Commission (Redecker et al., 2011). These trends can be easily connected to geomedia: (a) *personalization of learning* results from the student-centred approach of teaching, as geomedia enable the design of lessons based on the active participation of students; (b) *collaborative learning* occurs when students are engaged in team work and networking; recent technical progress facilitates the design of collaborative geomedia that enable such an approach; (c) *informalization of learning* results from the opportunity to learn anywhere at any time. Consequently, with geomedia embedded in mobile devices, learning is no longer limited to the narrow space of the classroom. This *mobile* or *ubiquitous learning* allows a rethinking of the value of *fieldwork* in education (Sanchez, 2008).

The readers of this book will have the opportunity to explore numerous examples illustrating and applying such trends. This collection will hopefully provide both a rich source to inspire everyday education and starting points for creativity. With this, their students will both experience learning supported by geomedia and be empowered to utilize geomedia maturely in their own everyday and professional lives.

References

Bachelard, G. (1938): *La Formation de l'esprit scientifique* (1992 edn). Paris: Vrin.

Baker, T. R., Kerski, J. J., Tu Huynh, N. & Viehrig, K. (2012): 'Call for an Agenda and Center for GIS Education Research'. *Review of International Geographical Education Online* 2 (3), 254–288.

Clement, J. (2000): 'Model Based Learning as a Key Research Area for Science Education'. *International Journal of Science Education* (22), 1041–1053.

Demirci, S. & Gryl. I. (2012): Mit dem Smartphone auf Spurensuche. Presentation at geo@web conference. Landau, 29 March 2012.

Dewey, J. (1938): *Experience and Education*. New York: Macmillan.

DiBiase, D., et al. (2006): *Geographic Information Science and Technology Body of Knowledge.* Washington, DC: Association of American Geographers.

DiBiase, D., et al. (2010): *Geographic Information Science and Technology Body of Knowledge* (2nd edn). Washington, DC: Association of American Geographers.

Donert, K. (ed.) (2010): *Using Geoinformation in European Geography Education.* Vol. IX. Rome: International Geographic Union – Home of Geography.

Döring, J. & Thielmann, T. (2009): 'Mediengeographie: Für eine Geomedienwissenschaft'. In J. Döring & T. Thielmann (eds.), *Mediengeographie: Theorie, Analyse, Diskussion.* Bielefeld: transcript, 9–64.

Douhei, M. (2011): *Pour un humanisme numérique: librairie du XXIème siècle*. Paris: Seuil.

Gruber, I: (2009): 'Fachübergreifende GI-basierte Lernobjekte zum Internationalen Polarjahr'. Masters thesis, University of Salzburg.

Gryl, I. Jekel, T. (2012): 'Recentering Geoinformation in Secondary Education. Toward a Spatial Citizenship Approach'. *Cartographica* 47 (1), 18–28.

Gryl, I., Jekel, T. & Donert, K. (2010): 'GI & Spatial Citizenship'. In T. Jekel, A. Koller, K. Donert & R. Vogler (eds.), *Learning with GI V*. Berlin: Wichmann, 2–11.

Hakley, M. (2010): 'The End of the "ArcInfo Driving Licence" Era'. *Cartographica* 45 (2), 85–88.

Halloun, I. A. (2004): *Modeling Theory in Science Education*. Dordrecht: Kluwer Academic Publishers.

Hegarty, M., Khooshabeh, P., Stull, A. T. & Dara-Abrams, D. (2011): *Hegarty Spatial Thinking Lab*. Available at http://www.psych.ucsb.edu/~hegarty/index.php.

Ingham, A. M., & Gilbert, J. K. (1991): 'The Use of Analogue Models by Students of Chemistry at Higher Education Level'. *International Journal of Science Education* 13, 193–202.

Jekel, T. (2007): '"What you all want is GIS2.0." Collaborative GI Based Learning Environments: Spatial Planning and Education'. In A. Car, G. Griesebner & J. Strobl (eds.), *GI-Crossroads @ GI-Forum*). Heidelberg: Wichmann, 84–89.

Jekel, T., Pernkopf, M.-L., & Hölbling, D., (2008): 'Rethinking Spatial Thinking. An Empirical Case Study and some Implications for GI-Based Learning'. in: K. Donert & G. Wall, *Future Prospects in Geography*. Liverpool: Hope University Press, 377–384.

Johannson, T. (ed.) (2006): *GISAS Project: Geographical Information Systems Applications for Schools*. Helsinki: University of Helsinki.

Kerski, J. (2003): 'The Implementation and Effectiveness of GIS in Secondary Education'. *Journal of Geography* 102 (3), 128–137.

Marsh, M., Golledge, R. & Battersby, S. E. (2007): 'Geospatial Concept Understanding and Recognition in G6 College Students: A Preliminary Argument for Minimal GIS'. *Annals of the Association of American Geographers* 97 (4), 696–712.

Milson, A., Kerski, J. J. & Demirci, A. (2012): *International Perspectives on GIS in Secondary Education*. Netherlands: Springer.

National Research Council (2006): *Learning to Think Spatially. GIS as a Support System in the K-12 Curriculum*. Washington, DC: National Academies Press.

National Science Foundation (2000): 'Inquiry: Thoughts, Views, and Strategies for the K-5 Classroom'. *National Science Foundation* (2).

Redecker, C., Leis, M., Leendertse, M., Punie, Y., Gijsbers, G., Kirschner, P. & Hoogveld, B. (2011): *The Future of Learning: Preparing for Change.* Seville: JRC-IPTS, European Commission.

Rocard, M., Csermely, P., Jorde, D., Lenzen, D., Walberg-Henriksson, H. & Hemmo, V. (2007): *Science Education Now: A Renewed Pedagogy for the Future of Europe*. European Commission.

Sanchez, E. (2008): 'Integrating Geomatics into the Curriculum: Inquiry-Based Learning During a Fieldwork Course'. In T. Jekel, A. Koller & K. Donert (eds.), *Learning with Geoinformation III.* Heidelberg: Wichmann. 78–87.

—. (2009): 'Innovative Teaching/Learning with Geotechnologies in Secondary Education'. In A. Tatnall & T. Jones (eds.), *Education and Technology for a better World*. Berlin: Springer. 65–74.

Sanchez, E. & Jouneau-Sion, C. (2009): 'Playing in the Classroom with a Virtual Globe for Geography Learning'. In T. Jekel, A. Koller & K. Donert (eds.), *Learning with Geoinformation IV.* Heidelberg: Wichmann. 78–86.

Schuurman, N. (2004): *GIS. A Short Introduction.* Oxford: Blackwell.

Steinmann, R., Häusler, E., Klettner, S., Schmidt, M. & Lin, Y. (2013): 'Gender Dimensions in UGC and VGI: A Desk-Based Study'. In T. Jekel, A. Car, J. Strobl & G. Griesebner (eds.), *GI-Forum 2013. Creating the GISociety*. Berlin/Offenbach: Wichmann & Vienna: Austrian Academies of Sciences Press. 355–364.

Schauppenlehner, T., Höglhammer, A., Muhar, A. & Eder, R. (2013): 'Rethinking Urban Space: GPS Drawing as a Tool to Examine Spatial Patterns and Relationships'. In T. Jekel, A. Car, J. Strobl & G. Griesebner (eds.), *GI_Forum 2013. Creating the GISociety*. Berlin/Offenbach: Wichmann & Vienna: Austrian Academies of Sciences Press. 399–406.

Strobl, J. (2008): 'Digital Earth Brainware'. In J. Schiewe & U. Michel (eds.), *Geoinformatics Paves the Highway to Digital Earth (gi-reports@igf)*.Osnabrück: University Osnabrück.134–138.

Tu Huynh, N. (2008): 'Measuring and Developing Spatial Thinking Skills in Students'. In T. Jekel, K. Donert and A. Koller (eds.), *Learning with GI III*. Heidelberg: Wichmann. 116–125.

Van der Schee, J. & Horvath, G. (2006): 'Evaluating the GISAS Project: Constraints and Challenges'. In T. Johansson (ed.), *Geographical Information Systems Applications for Schools – GISAS*. Helsinki: University of Helsinki. 95–100.

Vogler, R., Hennig, S., Jekel, T. & Donert, K. (2012): 'Towards a Concept of "Spatially Enabled Learning"'. In T. Jekel, A. Car, J. Strobl & G. Griesebner (eds.), GI_Forum 2012 – Geovisualization, Society and Learning. Wichmann, Berlin. 204–211.

PART TWO:

LEARNING AND TEACHING WITH GEOMEDIA: SELECTED LEARNING ENVIRONMENTS

CHAPTER FOUR

TEACHING AND LEARNING WITH GEOCACHING

GIOVANNI DONADELLI AND LORENA ROCCA

Abstract

Is it possible to use technologies to transform teaching and learning in geography into an adventure, through the use of challenging games? The answer is most definitely 'yes', and what makes it possible is a recent version of the evergreen treasure hunt activity called geocaching. This is a game based on the use of GPS receivers to locate definite coordinates, and its educational value lies in the skills of the students and the actions they take to find the treasure. This paper briefly introduces the rules of geocaching, presents its educational benefits, and proposes three simple teaching activities aimed at promoting teaching and learning with geocaching at school.

1 Aim, Target Standard, Target Group

Geocaching is a modern treasure-hunting activity in which the participants look for hidden containers called 'caches' or 'geocaches' with the help of a GPS device. This engaging outdoor practice was played for the first time in America around the turn of the twenty-first century and today it is a global phenomenon involving over 5 million geocachers, searching for more than 2 million treasures, spread all around the world (www.geocaching.com, 23 April 2013). This huge community has its 'home' in the website www.geocaching.com, where a wealth of information about geocaching can be found, including details of the coordinates of caches and where all the new finds can be logged.

The essence of geocaching can be found in its name, which is formed by two parts: the first, *geo*, refers to the earth, geography and the position on the globe given by geographical coordinates; the second, *caching*, concerns the actions of hiding and finding containers.

The initial reactions of those who engage with geocaching are often a mix of curiosity and suspicion. The idea that someone has hidden little treasures in places people frequent daily (squares, parks or any other place in town), arouses curiosity. The aim of this paper is to prove that geocaching is not only a well-organized and pleasant game, but also an involving educational activity that, if taught effectively at school, can contribute positively to geographical learning.

Pedagogical Background. Geocaching offers challenging educational and geographical stimuli, and the teaching becomes a form of negotiation involving the active construction of meanings in which cooperation and collaboration among students are necessary. The processes of research and planning encouraged in geocaching offer students the opportunity to face challenging tasks, based on real and meaningful stimuli, which allow them to create, share, modify, predict, test and assess their own ideas about the space they will explore. Geocaching offers students graphic representations, written descriptions and field experiences to reflect on, either individually or in groups, encouraging them to consider processes related to the social construction of space and place. The approach is extremely inclusive and supports a variety of learning styles: linguistic, logical/mathematical, visual/spatial, interpersonal, kinaesthetic and digital.

Therefore, promoting geocaching at school essentially involves proposing 'experience-based' geography, renewed as regards its form but confirmed in its substance, motivating the students, and developing out of their daily experiences.

Moreover, geocaching has the merit of creating a context of meaning in which it is possible to invite the students (even as young as 8 years old) to discover and to learn the use and the potentialities of digital cartography and GPS devices. The use of technology is playing an increasingly important role in geography, allowing some operations which otherwise would be very complex and time consuming to be carried out simply and rapidly. Nevertheless, geographical knowledge is still crucial because spatial analysis depends on data as well as on the ability to interpret the data properly. Geocaching actively supports the development of such capabilities. The GPS device helps to identify the place where the treasure is hidden, but it does not give the exact location. In geocaching, technology is therefore presented as a useful tool, but it is still a medium, which contributes to building scaffolding for learning. That is why we propose an experience 'with technologies' rather than 'about technologies'.

Aims. The aim of the teaching activities we propose is to motivate and guide students in learning geography through short activities inspired by geocaching. In particular, these activities are meant to enlarge and consolidate the vocabulary of geography, to strengthen the competences linked to the students' personal orientation, and to promote their spatial thinking through geographical sources (digital and non-digital), analysis, hypothesis/abstraction and reflection. A further objective is to contribute to the creation of a 'personal' and respectful relationship between the student and the environment (Taylor, 2010, 18) by promoting the use of an investigatory approach, supported by the introduction of technologies such as the GPS device and online cartography services. Through these activities and their discussions, students will be encouraged and stimulated to think geographically and to know where, how and why to think spatially (National Research Council, 2006).

Finally, this contribution offers teachers a cross-curricular teaching tool, with opportunities to expand learning in language, mathematics, science and social studies areas, highlighting the central value of geographical knowledge as a link with other subjects.

Target groups. The activities described in the following paragraphs are designed for 8- to 18-year-old students of all abilities. No particular technological skill is required beforehand.

2 Description of the Learning Environment

Geocaching is an activity that is most effectively undertaken outdoors. Being outside contributes to the success of geocaching and, even if it is possible to carry out some indoor activities (see step one: pre-geocaching), the opportunity to enable students to experience learning outdoors is compelling. It is possible to play geocaching individually or with a classmate. At school, the tendency is to have the students work together, in order to obtain good team work during the activities; it is important to give the students the opportunity to practise doing tasks and taking decisions in small groups.

Figure 4-1: Different kinds of geocaches

The resources needed by the teacher to introduce geocaching at school are: a compass, a GPS receiver, a computer connected to the Internet, a container, and an item to be hidden. There are two features that turn a container into a 'geocache' (Fig. 4-1). Firstly, a geocache always has to have a logbook, or a little notebook, or a piece of paper, where the finders can 'log in' by writing their name. Secondly, it is necessary to have the geographical coordinates of the points where the geocaches are hidden, as they have to be given to the geocachers. Following these two simple rules it is possible to create geocaches of every material and shape, although the geocaches have to be camouflaged in the environment in order not to draw the attention of 'Muggles', a word taken from the Harry Potter books: in the geocachers' language, Muggles are all those people who do not play geocaching and who could interfere with the location of the caches. A geocache can be made from a food container, or, for example, chocolate egg surprise containers, which are small, easy to acquire, to camouflage and to hide.

Obviously, the higher the number of items hidden, the greater the number of students who will be able to participate at the same time.

3 Preparation for the Teacher

Three different geocaching activities are suggested, offering increasing levels of complexity, to develop students' capabilities. These are supported by a list of the materials needed, some simple variations, and a list of further resources. The activities can be considered sequentially, but they can also be taken in a different order, or as a single experience.

Step 1: Pre-Geocaching. The first activity is a simplified 'hunt', allowing the students to familiarize themselves with the technologies and the logic of geocaching.

Step 2: Edu-Geocaching. The second activity is a simulation that proposes the discovery and the positioning of the caches as a teaching strategy to enhance geographical understanding and spatial thinking.

Step 3: Go-Geocaching. The third activity involves researching, planning and finding the first official geocache. The search takes place in groups, and it will enable students to develop their own geocaching skills beyond school.

The inspiration comes from some school experiences realized in 2012 in Italy with 8- to 18-year-old students. It should be noted, however, that even though the activities offer detailed instructions, they will need to be adapted to suit the potential of the class as a whole and the abilities of individual students. This is necessary in order to obtain a good result from the experience, but also to develop the sought-after 'Community of practice', which is an ideal condition for developing competence.

Pre-Geocaching

Specific aims: define common terms for map and compass; consolidate knowledge and skills as regards cardinal points; practise how to use a compass to determine directions of travel; introduce what a GPS unit is and practise using it; promote the creation of mental maps; enhance students' personal geographies.

Materials: for each group: 1 compass, 1 GPS device, 1 container to use as a cache; 1 map of the areas outside the school, drawn on a poster. A guide to buying a GPS can be found here: http://www.geocaching.com/about/buying.aspx. Before buying one, teachers might like to check whether they can borrow a GPS device from the community (university, associations, research centres, etc.).

Spaces and duration: the activity is best carried out in the school grounds (or, if these are too small, in an open space near the school); it takes about 90 minutes.

Prerequisites: before starting this activity, check students' knowledge of the points of the compass and their ability to use a compass to orientate themselves and move in the appropriate direction.

Instruct students to take 20 steps in the direction of 90° (i.e. eastward); instruct them to orientate the map of the school again correctly (i.e. with north at the top); read the compass bearing and give the classmates the direction by using degrees derived from the compass.

First part: hunting

Divide the class into teams, and give each team a written clue. This clue should guide them to where the hunt will start. By following the clues every team should find a camouflaged box (the initial cache), which will contain a compass, a GPS device and at least 5 clues to reach the treasure (last cache). Every clue contains the direction in degrees (called a bearing or azimuth; more information about this can be found in the paragraph 'Further resources') and the distance to cover (e.g. *move forward 43 metres in direction 180° N*). For each section, the members of the team should exchange roles (listed below), in order to give everyone the opportunity to test their skills (Fig. 4-2).

The roles are:

Leader: this student holds the clue, gives his/her classmates instructions and has responsibility for the team's movements;

Student responsible for the compass: this student carries the compass and chooses the direction to follow;

Student responsible for the distances: this student measures out the metrers the team needs to cover with his/her steps;

Student responsible for the GPS: this student needs to understand what the GPS is used for and how it works;

Observer: this student writes down observations of phenomena at the location of each clue.

After following all the clues, every team should reach the cache. Inside the cache, the students will find a small sheet of paper, providing them with a few simple instructions for how to retrieve the coordinates of a point using the GPS device, and several tasks. They then write down the coordinates of the cache, the waypoints where they have changed direction, the starting point, and an additional point, such as a landmark representing the school campus (e.g. the centre of the football field).

Figure 4-2: Each student needs to have a role in the team

Second part: debriefing

The students share their learning, including impressions and emotional responses. After the comments on the experience, the teacher should guide the conversation to the GPS device and its use and try to bring out the learning developed during the first part of the activity. Next, the teacher places the map of the school spaces on the ground, and asks every group to mark its waypoints, with their coordinates, on the map.

Once the map is completed, the students will see the coordinates of many points around the school and be able to discuss the meaning of those 'numbers'.

The activity is concluded by drawing students' attention to the fact that a set of coordinates identifies only one point on the earth's surface; it is not possible for two points with the same coordinates to exist (even if sometimes the GPS can be misleading).

Organization: The organization of this activity – i.e. to create and hide the caches and prepare the clues – could take up to 2 hours for a novice. This time can be reduced to about by half an hour if the preparation is done with colleagues. If it is left to only one teacher, the organization time can be reduced by doing a shorter activity (see the variations), or by following these guidelines:

'Plan before acting. Using a map of the activity area, identify up to 5 possible hiding places before deciding on the final location for the cache. In this way you will be sure to use all the space, and you will already have pre-chosen the locations of the materials.'

'Devise short written instructions and descriptions of the roles in order to make the task clear and understandable for the students.'

Assessment suggestions: Orientate the final map (with the marked points) and use it to enable your students to calculate the bearings of the movements made by the other teams. Ask every team to provide new clues to find their final cache; have them checked by another team. Organize a final discussion in which the teams can compare and assess each other's work.

Variations: If you do not have enough time to do this activity in its entirety, give the students the clues, compasses and GPS device at the start rather than have them search for the 'initial cache'. Alternatively, if you want to enrich the activity, you can ask each team to go and search for the other 'final caches' too, retracing the entire course with the compass or using the waypoints written on the poster.

Further resources: In order to consolidate the understanding of compass bearings, it is possible to organize a range of games using compasses (see e.g.: http://www.barossalight.sa.scouts.com.au/games/ compass-games.pdf or http://dragon.sleepdeprived.ca/games/compass/ compass.htm). Another very simple activity using compasses and GPS is 'Where in the World?', proposed in *Geocaching for School and Communities*, by Taylor et al. (2012, 122–129).

Edu-Geocaching

Specific aims. Use a GPS device to individuate waypoints of tracks, promote spatial thinking in the school environment, and support geographical teaching and learning.

Materials. 1 GPS and 1 'normal size' container for each group; 6 additional containers, and at least1 tape measure at least a metre long.

Spaces and duration. It would be better to do this activity in the school grounds (or, if they are too small, in an open space near the school); the activity takes about 2 hours.

Prerequisites. Students need to be able to use a GPS receiver to search for, reach and save the waypoints.

Activity to check the prerequisites: put the coordinates of a point in the GPS; follow the GPS instructions in order to reach the point; take 25 steps eastwards; save the waypoint as 'A1'.

Description: Before the activity starts, divide the students into groups. Each group will have a GPS: they will have to activate the 'track' modality on the GPS device and to hunt for 5 caches in the order given by the teacher (different for each team). In each cache they will find a different instruction, for example:

'Draw a map that includes the following: the school, the caches you have found so far, and other key features which, in your opinion, it is important to map. Then use colour to identify different areas.'

'Write down the categories of people who might find the cache (students, teachers, gardeners, parents, etc.). Explain how they might come across it.'

'Estimate and note down the distance in metres from the position in which you are to some key features nearby (e.g. a building, a tree, a fountain …). Then use the tape measure from the cache to measure the real distances to at least 3 of your chosen features. Make a note of the actual distances and suggest reasons for the differences between your estimate and the real distance.'

'Consider whether the place where you found the cache has always been as you see it now. What changes may it have been subjected to and by whom? Write down your thoughts and discuss with your group.'

'School spaces are used in different ways. Pick at least 3 spaces that have more than one use. Who decides how to use the school spaces? Are you happy with how the spaces are used? Do you have any suggestions for alternative uses? Write down your thoughts.'

It is important that all the students have followed the instructions and carried out the task before going on to find the next cache. Only those who have completed the assignment are allowed to sign the logbook in the cache. Once every member has signed the logbook, the team can proceed and look for the second cache, and so on.

Once all the students of a team have written their name in all the 5 caches, they can go and look for the final cache. Inside it, they will find the last instruction, which will have to be carried out by the whole team:

'Go to your class and take an empty cache. Hide it in a very special place, in the school or school grounds. Choose the place together as a group and write down its coordinates. Think about a possible task linked to that place, write it down, and leave instructions for the task in the cache for your classmates to find and carry out. For example, you can ask them to carry out an exploration or an investigation, take measurements, consider ideas, make a plan, write a presentation, or carry out some other activity that highlights the features of that place.'

When every group has prepared and located its cache, the groups exchange the coordinates. In this way each group will receive a different waypoint to reach. The activity ends with a group discussion. As for the previous activity, it is possible to reduce the preparation time by sharing it with a colleague. Hide the caches in isolated places from which it is not possible to see the other caches. Staple the instructions in groups of 5, and organize cache preparation in the classroom to allow students to be more creative using schoolbooks, atlases and/or computers.

Assessment suggestions. Prepare a simple evaluation grid (cooperation, patience, mutual respect, loyalty) in order to observe the groups' behaviour during the activity. Ensure the students assess both the contents and the creativity of the cache made and hidden by each team. The evaluation will be summative and take into consideration the participation in the team and how the instructions have been carried out individually.

Variations. This activity is structured so that you can carry it out several times with different aims. It will be sufficient to identify new hiding places and to change the instructions in the caches in order to update the adventure.

When the students have improved their skills, you will be able to add intermediate waypoints and to reduce the size of the cache. You can also encourage the students to organize and carry out the whole activity in groups, starting from the knowledge acquired on a particular topic.

According to the nature of the challenges put in individual caches, it will be possible not only to study a topic in depth (e.g. local history), but also to develop other skills, such as mathematical ones.

Further resources:
20 additional practical ideas for furthering students' work and take advantage of geocaching for school learning, accompanied by Field Sheets and Teacher Templates, can be found in the handbook 'Educaching – A GPS Based Curriculum for Teachers', published by SDG Creations Ltd. (2010).

Go-Geocaching

Specific aims. Promote planning skills, encourage 'informed spatial thinking', understand and investigate spatial representation such as digital maps and satellite imagery, introduce geocaching as a lifelong learning activity.

Materials. 1 GPS device for each group, an ICT suite with Internet access and Google Earth installed on every computer (www.earth.google.com).

Spaces and duration. The activity starts and ends in the ICT suite, but will also take place outside the school, near one or more caches selected from the web.

Prerequisites. It is necessary to be registered with the website www.geocaching.com (free) and to be able to use a GPS device to search for, reach and save the waypoints.

Fieldwork activities offer the teachers the opportunity for geocaching outside the school premises. Preparation should start some days before the field day where the teacher introduces the idea of geocaching (i.e. as an outdoor recreational activity and an international movement, in which the participants use a GPS device and navigational skills to hide and look for containers, called 'caches'). Students then explore the map on which all the 'official' caches are displayed (www.geocaching.com/map). The students will have to choose the ones which they believe can be found during the school trip. Later, they will be guided to gain information and to think about the places where the chosen caches are. Making use of digital maps and geo-referenced images (OpenStreetMap; Google Maps, StreetView and Earth; Bing Maps; Yahoo Maps, etc.) as well as other web resources (Wikipedia, web pages, blogs, pictures, webcams, etc.), students amass not only historical references or geographical features of the area, but also the approximate distances, the expectations and the perceptions linked to the places, possible hiding places, and difficulties that may be encountered when looking for the caches. At the end of the discussion, the waypoints of the selected caches will be saved on the GPS.

On the day of the trip, divide the students into groups and give each group a GPS to enable them to search for the cache. Students do not take the containers once they have found them: they simply let their teacher know that they have located a cache but they leave it in place for other groups to find. Once all the caches have been found or the time is up, the students are allowed to open the caches, to see what is inside and sign the logbook. Before the end of the activity, the students must put the caches back in exactly the same places where they found them.

When the students are back at school, instruct them to visit the website www.geocaching.com and log their findings (the students might also decide to log the findings individually, by using their smartphones). Finally, as a means of consolidating their experience and learning, invite the students to identify, using digital cartography, a cache that they would like to look for on their own. They will have to find information about the place in which the cache is hidden and to make a presentation combining cartography, satellite images and historical information based on the actual location for the hidden cache. This task will hopefully be exciting, ensure that the students practise their spatial thinking, and additionally motivate them to go geocaching outside school.

Before starting the activity, check whether there are any caches along the route to the start and, if possible, check their condition in advance. Being aware of where the caches are in advance will make it easier to run the activity. In order to be able to see the exact coordinates of the caches, it is necessary to be registered on the website www.geocaching.com. It is also possible to see the caches using Google Earth by downloading the 'Geocache Google Earth Viewer' (http://www.geocaching.com/about/google.aspx).

Assessment suggestions. Provide a form on which the students can record what they think they will find near the cache. Once the search is complete, students reflect on and correct their initial notes. It is also possible to provide a grid in order to evaluate the presentations prepared by the other students. Finally, the students should be tasked with evaluating the completeness and the accuracy of the information, the originality of the approach, and the ability to combine different points of view in the description of the environment.

Variations. If there is more than one cache near the school, you can give the students some waypoints and specific requests as homework. The students will have to find the cache and complete the assignment required. They could also create and place an official geocache near the school, or in a place connected to the activity done in class. An introductory video about geocaching, to explain the approach to students, can be found at http://www.geocaching.com/videos/. Many other ideas about how to take advantage of geocaching in teaching can be found in 'Geocaching and Education Subforums': http://forums.groundspeak.com/.

4 Experiences and Conclusion

The activities described in this contribution offer guidelines for using geocaching as an educational and teaching strategy at school. Experiments carried out by the authors in primary and secondary schools in Italy to test its feasibility proved to be very effective. All the students involved appreciated the activities, in particular the edu-caching, as this was carried out autonomously and was perceived as more active and involving than the other two activities. The main strengths underlined by the students were the cache search phase, the use of a GPS, digital cartography, and group work.

> 'What I liked most of all was to have to look for different caches; as soon as we found one, I immediately started thinking about where we could find the next one.' (Camilla, 8 years old)

From the teachers' perspective, what impressed most was the students' participation, the potentialities of geocaching as a geographical tool, the flexibility and the cross-curricular nature of the activities.

> 'Through these activities I succeeded in making the students reflect on the different uses of space and on their value, also stimulating some fruitful discussions […]. Geocaching has been a good opportunity to experience fieldwork and to show the stimulating side of geography.' (Massimo, teacher of 11- to 14-year-old pupils)

In fact, all the positive aspects which have been underlined are closely inter-related. The active participation of the students was probably due to the fact that they enjoyed looking for caches using GPS and digital cartography; these two factors strongly contributed to the students' motivation and, in turn, the students did not hesitate to become completely involved during and after the conclusion of the activities. They worked in groups, and it was often possible to hear the reflections they expressed aloud, which contributed to increasing the pleasure of working together and the ability to take joint decisions. Moreover, the students experienced that group work ends only when all the members of the group have finished, and that 'doing group work' is far more pleasant and fruitful than simply 'working in a group'.

Thanks to the way the activities were organized, the students were able to make and test their hypotheses, to ask questions, and to look for specific answers through a circular process of reflection and action that was shared in the group. Exploring and observing school spaces were useful exercises to practise taking a geographical view and engaging in spatial thinking. Using a compass and GPS in real contexts helped them to understand the functions of the devices and to practise personal orientation. These activities also led the students to exchange their views on different types of representation (the plan, the map) and to create their own maps, after reflecting on what they wanted to represent and the best way to do it. The three activities, in different ways, encouraged creative thinking by stimulating each student to find alternative solutions to problems, contribute to discussions, and to get involved in the group by making hypotheses, estimating distances and helping classmates. Moreover, the reflection on space and the places associated with daily life helped the teachers to guide the students to processes of abstraction, leading to the creation of mental maps (*'How can we reach a certain place?'* or *'Which way did we go?'*) or to an increased awareness of particular situations in their town observed during field experiences (e.g. affluent areas, public parks with closing hours, etc.). The fact that the students lived this experience as protagonists led them to pay attention to the places where they live, to ask questions about them (and to question themselves) and, as a consequence, to understand the value of space and of spatial thinking. The activities described above allowed the students to realize that geocaching is not only about searching for the cache starting from its coordinates; it also requires planning, strategy and spatial thinking.

As far as the above activities are concerned, it is necessary, however, to point to two weaknesses that emerged. The first concerns time, as the duration of the activities as suggested does not include time to discuss the ideas that may arise. The second is the availability of technologies. In this case, all the schools had to borrow GPS devices from universities or associations, which set limits to the organization and the management of the activities.

Nevertheless, these activities prove that it is possible to teach geocaching at school and that, through it, learning and adventure can coexist.

> 'All the energies that I had spent in the organization of these lessons were fully repaid by the enthusiasm with which my students participated in the following geography lessons.' (Ester, teacher of 9- to 11-year-old pupils)

The most rapid and effective way to discover the educational value of geocaching is to test it personally – a teacher more often enthusiastically promotes ideas and activities that he or she has experienced with pleasure. We invite you, our readers, to search for the nearest treasure (you can find it at www.geocaching.com/map) and to become involved in the game yourselves.

Acknowledgements

Our thanks go to all the schools and all the teachers who collaborated in this project for their participation and their helpful comments. In particular, we would like to thank all the staff from the School Institute 'G.A. Farina' in Venice for their invaluable support.

References

Gillin, P. & Gillin, D. (2010): *The Joy of Geocaching: How to Find Health, Happiness and Creative Energy Through a Worldwide Treasure Hunt*. Fresno, CA: Quill Driver Books.

Lo, B. & International Society for Technology in Education (2010): *GPS and Geocaching in Education*. Eugene, OR: International Society for Technology in Education.

National Research Council (2006). *Learning to Think Spatially*. Washington, DC: National Academies Press.

Taylor, J. K., Kremer, D., Pebworth, K. & Werner, P. (2010): *Geocaching for Schools and Communities*. Champaign, IL: Human Kinetics.

Simmons, B. & J. Seva (Eds) (2010): *The Complete Idiot's Guide to Geocaching*. 3rd edn. Indianapolis, IN: Alpha Books.

Web

Geocaching.com

CHAPTER FIVE

COLLECTING GEO-DATA TO SUPPORT CLASSROOM FIELD STUDIES

JOSEPH J. KERSKI AND THOMAS R. BAKER

Abstract

Converging societal, technological and educational forces are enabling the collection of geo-data to support classroom field studies. These forces also include renewed emphasis on eco-literacy, Science, Technology, Engineering and Mathematics education, career and technology-focused education, the migration of GIS to the web, calls for increased outdoor education, and increased interest in spatial thinking across the curriculum, inquiry, critical thinking, and problem-based education. We contend that these forces could represent the tipping point that greatly expands the use of geotechnologies in education.

In this chapter, we feature three rich environmental field study experiences that combine mobile devices and technologies such as web-based GIS, GPS, probes, cameras and smartphones to support rich, deep scientific and geographic inquiry. Web-based GIS such as ArcGIS Online enables the analysis of field-collected data to create visualizations. These tools encourage student decision-making, hypothesis-testing, and the formation of conclusions. The three case studies – a lichen project in formal education, a macro-invertebrate field experience as part of a species documentation project in informal education, and the gathering of field data with a live web-mapping service in a professional development workshop for teachers – illustrate the expanding possibilities that result from these converging forces.

1 Introduction

The US Center for Ecoliteracy's six core ecological concepts – networks, nested systems, cycles, flows, development and dynamic balance – create a context for thinking about and using geotechnologies in almost any classroom. GIS, as the flagship geotechnology, allows variables to be input, modelled and modified so that the dynamics of environmental processes can be studied. Hungerford and Volk (1998) defined nine key ecological concepts necessary for environmental education programmes: individuals and populations, interactions and interdependence, environmental influences and limiting factors, energy flow and nutrient cycling, community and ecosystem concepts, homeostasis, succession, humans as members of ecosystems, and ecological implications of human activities and communities. These topics are foundational to any project-centred GIS student work and should be discussed as students identify problems, design sampling protocols, collect data, and begin to answer questions of environmental and geographic significance.

GIS can enhance the teaching of these concepts. The North American Association for Environmental Education defined environmental literacy as including four interrelated components: competencies, knowledge, dispositions, and environmentally responsible behaviour. By using the same tools as scientists, GIS aids in the first two of these, and by investigating real issues in students' communities and beyond, it aids in helping with the last two of these components.

Students who use GIS in tandem with environmental studies develop key critical thinking skills. These skills include understanding how to carefully evaluate and use data (Bodzin, 2010). This is especially critical in assessing geographic data, due to its increasing volume and diversity, and given its often sensitive and politically charged nature. Geotechnologies were created and are continually being developed specifically to solve problems. As such, they are key tools used not only by students, but also by hundreds of thousands of practising scientists, planners, analysts and decision makers around the world on a daily basis.

In addition to advances in environmental education, several converging technological forces have enabled educators and their students to easily collect environmental, social, geological and other data in the field. More importantly, they have enabled students to subsequently map those data in a spatial context. The data may cease to be simply location-based, then, and become important in analysis and decision-making. This can happen. But is it in fact happening? Three case studies illustrate that this is not only happening, but that it is doing so to varying degrees, at different levels of education, and in very varied content areas.

2 Mapping Geotagged Lichen Images

On a crisp winter's morning, 26 seventh graders briskly move about the school grounds of Wheatridge Middle School, in Gardner, Kansas, USA. Students in groups of four and five are bouncing from tree to tree with acetate paper, log book, digital camera, and GPS receiver in hand. They have been tasked with creating a quick biological inventory and map of lichen growth and damage on tree bark. Using the acetate grid, students quickly measure and record lichen coverage around the trunk of a tree, take a quick photo, and record the GPS receiver's latitude and longitude. Groups race to record at least half a dozen trees in the allotted twenty minutes outdoors before running back inside, where they will begin to compile their data in a spreadsheet.

While collecting lichen data, students record a variety of other relevant data such as tree species, date and time, and location information with a GPS receiver. Students bring this data back to school where it can be entered into a spreadsheet and then plotted in a web-based GIS, such as ArcGIS.com, to see trends in lichen health. A unit of study like this often takes two to three weeks, involving classroom and computer lab time, as well as data collection at home. This particular project is described in the 'Lichens and SO_2' project (e.g. http://edgis.org/so2).

During data analysis, students create a map of lichen coverage from the data collected. Like any other attribute recorded during a data collection event, pictures of the phenomena are often highly valuable in geography and tell a story of their own. Photos are first geotagged and then mapped. Geotagging is the process of adding locational information to an image. Consider a student inventory of trees, birds, fossils, or virtually any phenomenon that occurs in the natural world and occupies a location. These natural inventories require data collection and if 'a picture paints a thousand words', the quantity and reliability of student-collected data can be aided greatly. Whether looking for trends and patterns in field data or simply creating a map-based field guide, geotagging offers a unique perspective on viewing data.

Most smartphones and many new cameras automatically embed the location in the photo's metadata. For those with other cameras or phones, a variety of free desktop or web-based photo editing and sharing software can easily geotag an image. By geotagging an image, we know where the image was acquired and can place it on a map. This process is made most efficient when students use a GPS receiver to capture the latitude and longitude and record the data in a log book.

Geotagging images require desktop software, such as current versions of Adobe Photoshop Elements, Google Picasa, or Apple iPhoto, which are free or low cost. Specific steps for geotagging vary across software, but all require just a few clicks (e.g. http://edgis.org/geotag). Most geotagging software uses an Internet connection to contact a server that converts addresses or place names to coordinates of latitude and longitude. After the software has geo-tagged the image, it is ready to be mapped.

The geo-tagged images are mapped by placing the image on the map or digital globe. Students immediately see how density and diversity of characteristic lichen damage can vary regionally. Additional numerical data can then lend support to each point as students examine in depth each sampling event. Flickr and Google Picasa are excellent destinations for publishing students' photos of lichens if the school allows photo sharing. Alternatively, students can create geo-tagged maps on 3D globes using desktop software like Esri ArcGIS Explorer (http://esri.com/arcgis explorer) or Google Earth (the latter also currently requires Google Picasa). These globes allow the user to see geo-tagged photos, properly positioned on the planet, while supporting additional content such as GPS log files, varied base maps, and even data from GIS.

Using these same techniques, students could photograph leaves, rocks, trees or animals around their community. They could just as easily photograph birds, butterflies or insects, depending on the subject matter being taught. When students begin taking geo-tagged pictures in their own neighbourhoods, sharing those images with peers and then looking for patterns, the educator is encouraging students to develop higher visual learning and geographic skills.

3 Collecting, Documenting and Mapping Species Data

In an alpine wetland in Rocky Mountain National Park in Colorado USA, students aged 11 to 13 collect macro-invertebrate data in a two-day effort to document as many species as possible within the park's boundaries. As expected, students are interested in collecting the macro-invertebrates with poles, nets and trays, getting wet and muddy in the process. But they are even more engaged in classifying the macro-invertebrates. In fact, one group is so engaged that they do not notice a bear gazing at them from 100 metres away! Data are recorded manually on paper forms that were created according to a standard protocol, containing latitude, longitude, taxonomic category, macro-invertebrate scientific name, and common name, for each sample collected. Next, back at the school, the data from the paper forms are entered via laptops and tablets into a spreadsheet, saved in a comma-separated value (CSV) format, and then mapped via ArcGIS Online (http://www.arcgis.com/home), a web-based mapping platform, allowing students and educators to create, store and share maps and applications in the cloud. Inside ArcGIS Online, the spreadsheet is added via an 'Add layer from file' function. Each data row in the spreadsheet becomes a point on the map, with the attributes collected in the field. The students enable editing on their points, which allows them to link each point to their own photographs, scanned sketches or videos that they have stored online, for example in Dropbox, Picasa Web, Flickr, or on their own web pages.

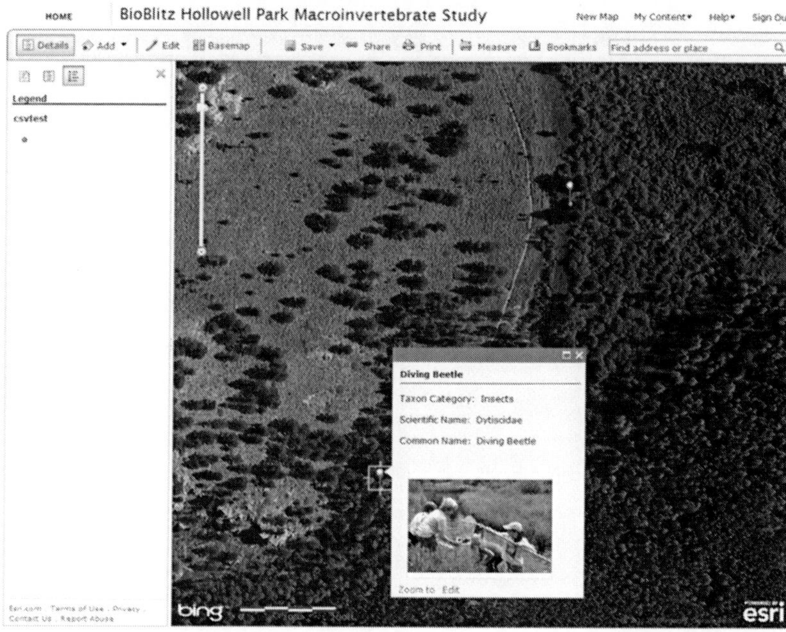

Figure 5-1: BioBlitz.

Once the data have been entered into ArcGIS Online, the map is saved into the class's ArcGIS Online account, and shared with everyone via a URL. The map is also embedded into the class's section of their school's website. The students embed the map by copying the embed code that is generated by ArcGIS Online and pasting it into the HTML document that runs the class's web page. The student-collected data, including the measurements, images and field notes, are now accessible to fellow students, teachers, park scientists and the general public to examine. Students are required to write a report and give an oral presentation about the health of the wetland, using their data to back up their statements. As they write their report and prepare their presentations, students change the base maps behind the data to a satellite image, and to topographic or other maps in order to help them understand their study site and its broader geographical context. Thematic layers, such as information on water quality and quantity, a geological map and ecoregions of the park, are also added. Most importantly, the students can see their data in context, comparing their data to other macro-invertebrate data in this location and in other locations collected by other students that day. Students then assess the overall ecological diversity of this wetland and how it compares to other wetlands at similar elevations in the park. The students are required to use the map to which they contributed in their presentation, but this part does not seem forced, as the map has already become a platform for telling their story. In fact, thhe map has now become a multimedia platform. Furthermore, if the teacher takes the students to this same location in subsequent years, data can be added to the existing spreadsheet and map, allowing students to assess how this wetland changes over time.

4 Crowd Sourcing Fieldwork: Collecting and Mapping Data in a Collaborative Web-GIS Environment

A teacher stands next to a tree and touches a screen on a smartphone, adding her location information in the process. She then uses a drop-down menu to enter the height, species and condition of that tree. Her iOS-based smartphone is running the ArcGIS Online app, which allows her to immediately see the point that she just added on a satellite image basemap. Not far away, a group of her fellow teachers do the same thing for other trees on the school property, using a combination of Android and iPhone devices. Each teacher not only sees the points that he or she collects, but also takes note that the other points now appearing on the map in real time are from his/her peers. Thus, the teachers are collectively adding to a single online map. This map can be updated during the entire semester-long course, and added to in subsequent years, resulting in a long-term map-based story of the ecology of the park, the riparian zone, or the school campus.

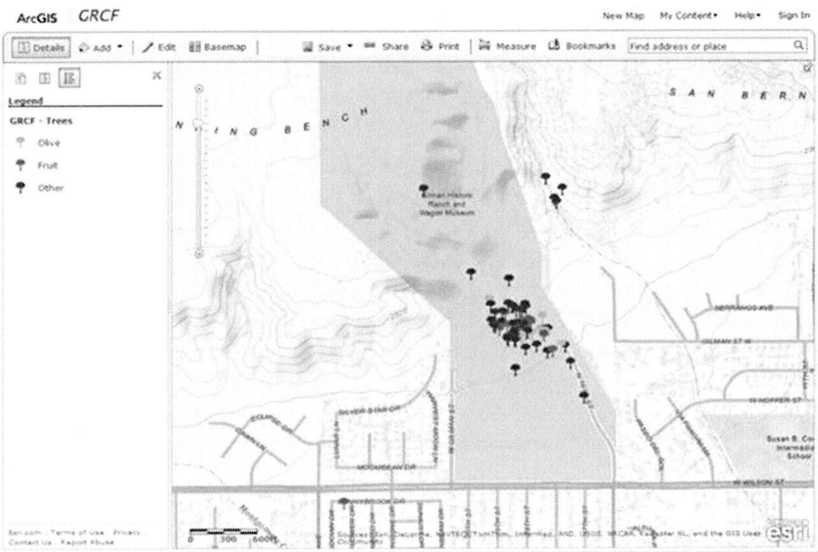

Figure 5-2: Example of the ArcGIS Online app.

How was the above field experience made possible? Before entering the field, an editable map was created in ArcGIS Online for the purpose of mapping tree locations and three characteristics of each tree. The instructors created a simple geo-database, using ArcGIS for Desktop, that contained ranges for each accepted value. These ranges include acceptable values for the trees' height in relation to specific tree species, and condition (excellent, good, fair and poor). These values became the drop-down menu choices that the teachers can see in the field on their smartphones. The geo-database was turned into a feature service online from ArcGIS for Desktop. Once online, it was saved as a map, shared with everyone, and made editable. It could then receive input from the teachers from smartphones in the field, or via tablets and laptop computers in the school building or at any location. Though easy to set up, this capability has far-reaching ramifications when used with students. A similar editable feature service could be created for national or international surveys of student data, or to collect environmental or social data, such as water quality measures of local streams and lakes, the amount of litter in a 10 square metre area outside entrances of different schools, the distance from home to school, or any other mappable data. Students can set up the editable maps so that they have input over the entire process from start to finish. Integrating these editable feature services into education provides not only meaningful field experiences in which students can collect data using ordinary low-cost devices, but also a way for students to analyse their data spatially.

5 The Crucible: A Look at Societal, Educational and Technological Forces

Several forces are suddenly and irreversibly converging to enable the collection of geo-data to support classroom field studies. These forces may be grouped as *societal*, *educational* and *technological*.

Societal forces include a realization that tying a location to data is valuable for everyday decision-making in business (e.g. routing a fleet of vehicles), science (e.g. managing a national forest), in recreation (e.g. choosing which route to take while hiking), and in thousands of other ways. This is driving an unprecedented amount of development of hardware tools and software applications that are allowing everyone from ordinary citizens to policymakers to access data and tools that were previously available only to GIS analysts. This, in turn, has re-energized the citizen science movement, turning non-scientists into data collectors and analysers, roles that were formerly the realm of government agencies, private industry, and other organizations that were the chief content providers of data about the Earth.

Educational forces fostering the use of geotechnologies include a renewed emphasis on Science, Technology, Engineering and Mathematics (STEM) education, as well as career and technology-focused education. These forces are driven in part by a growing realization that scientific and technical skills are lacking in the workplace, creating a high demand and opportunities for graduates with these abilities. Another force is renewed emphasis on outdoor education. Calls for increased outdoor education (Louv, 2006) cite the dire consequences that can arise in a society where decisions are made by those who have had no field experience. Instructors who regularly employ field methods often do so with goals of fostering insight about processes, scale, the environment, enhanced engagement among students, holistic thinking, and tactile learning. While assessing the effectiveness of fieldwork is a challenge (Fuller et al., 2011), it is clear that students can increasingly integrate technological tools into their field experiences, thanks to the proliferation of mobile devices and web GIS.

In addition, calls for deeper inquiry, critical thinking, and problem- and project-based education are frequent in many curricular areas. Many of these emphases have at various points in the past become rallying points for education, prompting some to wonder whether innovation really is occurring or whether it is merely 'old wine in a new bottle' or 'lessons learned and relearned'. However, there is something new: two new, curious, and simultaneous forces now accompany renewals of old familiar emphases. First, there is a growing realization that GIS allows educators to achieve their goals more efficiently, rigorously, and according to national standards. GIS is increasingly seen as a technology that can help students develop critical thinking (Tsou & Yanow, 2010), and foster inquiry by allowing for the 'what if' questions to be examined, and it is something into which educators can embed authentic assessments. GIS is seen as a green technology and one that can help citizens become connected to their community and their world. Second, and consequently, there is increasing interest in spatial thinking across the curriculum at primary, secondary and university levels.

Technological forces include easy-to-use smartphones and their growing ubiquity, increased Internet bandwidth, and applications that tie data from numerous inputs into one system. This has provided classrooms with rapidly expanding access to devices that can collect digital, geo-locatable data and multimedia, fuelling a trend towards 'Bring Your Own Device' (BYOD) in schools. Using smartphones with built-in cameras and built-in GPS chips, students can record data as needed by their field study in 'citizen science' mode. Data are typically sent directly to cloud-based recording or visualization applications for immediate screening and interpretation, either by students or by teachers.

Yet for all of these developments, the migration of GIS to the web may represent the most important technological development of all. Students can map data in a web-based GIS such as ArcGIS Online by adding geo-data from GIS servers, local or remote files, or even by creating editable content directly inside the web-based GIS. The power and simplicity of newer technologies allow instructors to focus on pedagogy and the subject matter, making the geotechnologies increasingly transparent. This provides the means for student-collected data to be mapped, changing the perception of what a map is and what it can do from a static reference document to a powerful investigations and communications platform. Web GIS enables mapping and spatial analysis to be accomplished with a web browser. More importantly, it allows for results to be easily shared. In the past, sharing posed problems for many educators with desktop-based GIS. Lastly, the maps can be modified, embedded and queried. The longstanding encouragement from the GIS education community not just to collect data but to map it and analyse it is at last achievable. The era of 'GIS for everyone' may have begun.

We contend that the convergence of these forces could represent the tipping point that greatly expands the use of geotechnologies in education. We are beginning to see a wider adoption of GIS and spatial analysis by geography educators, but also by science, mathematics, history, language, arts and technology educators. Could this adoption lead many educators to cross the 'adoption chasm', and thereby open up these technologies and methods to all 'early adopters' and even the 'early majority' of educators? Early indications seem to be pointing to this eventuality.

6 Conclusion

Why does all of this matter? Students who are well grounded in the geographic perspective through geotechnologies are better able to use data at a variety of levels, in a variety of contexts, think systematically and holistically, and use quantitative and qualitative approaches to solve problems. In short, as graduates, they are better decision-makers. Students engaged in GIS make heavy use of the geographic inquiry process, which involves asking geographic questions, acquiring geographic resources and data, analysing geographic data, assessing the data, and making decisions from the resulting geographic information. This leads to additional questions and inquiry, continuing the cycle. Environmental literacy will be fostered if students are provided with solid content knowledge, skills and competencies, combined with the spatial framework that working with these tools allows.

If these forces convince even just the early majority of educators to embrace geotechnologies, it could transform education. How? Teaching with these tools and methods is based on inquiry and problem-based learning. Inquiry will expand in education through spatial analysis and geotechnologies. Furthermore, with geo-enabled graduates in key decision-making roles, societies will become more efficient and sustainable.

For Further Exploration

The Esri GIS Education blog is published several times each week, focusing on how and why to use GIS in education (http://edcommunity .esri.com). Maps can be easily made and shared with ArcGIS Online (www.arcgis.com/home).

References

Bodzin, A. (2010): 'Integrating Web-Based Activities and Field-Based Experiences to Investigate Environmental Issues'. In A. M. Bodzin, B. S. Klein & S. Weaver (eds.), *The Inclusion of Environmental Education in Science Teacher Education.* Dordrecht: Springer, 323–336.

Fuller, I., Edmondson, S., France, D., Higgitt, D. & Ratinen, I. (2011): 'International Perspectives on the Effectiveness of Geography Fieldwork for Learning'. *Journal of Geography in Higher Education* 30 (1), 89–101.

Hungerford, H. & Volk, T. (1998): *Curriculum Development in Environmental Education for the Primary School: Challenges and Responsibilities. Essential Readings in Environmental Education.* Champaign, IL: Stipes.

Louv, R. (2006). *Last Child in the Woods.* Chapel Hill, NC: Algonquin Press.

Tsou, M.-H. & Yanow, K. (2010): 'Enhancing General Education with Geographic Information Science and Spatial Literacy'. *URISA Journal* 22 (10), 45–54.

CHAPTER SIX

LEARNING ABOUT WORLD AGRICULTURE WITH GOOGLE EARTH AND FAOSTAT

ÜLLE LIIBER, JÜRI ROOSAARE AND VAIKE ROOTSMAA

Abstract

Although compulsory themes of the geography curriculum like world agriculture and food production may not be particularly exciting for most secondary school students, we still use foodstuffs from all around the world on a daily basis. Academic and abstract learning materials in textbooks or traditional lectures tend not to be motivating study aids and methods for present-day digital learners. However, most students are technologically capable and therefore are inclined to enjoy learning through technology. Students can find a wide range of information on the Internet, but they often have problems with such things as critical analysis, making connections, generalization, argumentation and drawing conclusions relating to the data found. Even though there has been a lot of discussion about the importance of inquiry-based study, teachers often fail to implement it in their classes, because they do not have the necessary experience, the study itself is time-consuming, and state exams do not test the students' inquiry skills.

We compiled and piloted eight lessons about world agriculture for secondary schools. The lessons cover the required curriculum content and at the same time develop students' inquiry skills, spatial thinking, ICT literacy and collaborative skills. Students working in groups in computer classrooms investigate farming in different areas of the world using Google Earth, statistical databases from FAOSTAT, and other Internet sources. Instructions guide students to work independently with data, analyse, make comparisons and draw conclusions.

The results of the pilot showed that the students' interests in the topic increased, especially in working with Google Earth, when the students experienced the 'Aha!' effect. Because of the students' limited background knowledge on agriculture, interpreting satellite pictures from a farming perspective was not easy. Most students found all the necessary data from FAOSTAT and other sources with the help of an instruction guide; however, they had problems with comparison, analysis, reasoning and conclusions.

The digital earth.eu project could be a good platform to create a depository of kmz-files that would include relevant information on different topics of the curriculum, and is created collectively by teachers from different countries. Other geomedia resources could be added in the future.

1 Aim and Target Group

This project is aimed at students in class 11 (age 16) and is designed to cover 6–8 lessons of 45 minutes each, depending on students' capabilities. The students' first task is to describe different farming types in the world with the help of Google Earth. Students work in groups. Each team investigates agricultural landscapes of specific regions that are in different stages of development and located in different climate zones. They have to identify where the impact of agriculture on the physical environment is the greatest. On the basis of materials collected from different sources, the team then determines the role played by agriculture in the economies of the states concerned. Students then create team presentations that summarize and visualize information on slides.

The pedagogical aim of the project work is to develop each student's inquiry skills, spatial thinking, ICT literacy and collaborative skills, and knowledge and understanding of agriculture and its diversity and impact on the environment. The work also contributes to the formation of students' values and attitudes.

During the study, students' inquiry skills are further developed: students are required to ask geographical questions and find reasoning for their hypotheses. During lessons held in the computer lab, they collect relevant information from various sources such as databases, thematic maps, climate charts, photos, articles, and from other web-based sources. Students need to assess critically, interpret and analyse the data, find correlations, and make generalizations and conclusions based on the data collected.

They explore and interpret satellite pictures from Google Earth to describe the agricultural landscape. On the basis of the information collected, students recognize spatial relationships; describe and compare physical conditions (relief, climate, soils) for farming; identify the relationship between physical conditions, land use and agricultural specialization; analyse crop yield or productivity and relate it to the level of state development. Finally, they have to decide where the impact of agriculture on the physical environment is the greatest.

During this study, students' cooperation skills are encouraged and developed. Each team is required to plan the work, to decide on the distribution of subtasks and information, and how to support each other and formalize the results in group presentation. The more effectively team members communicate, the better the results.

One of the project's aims was to determine whether this study strategy justifies itself or not. For students, this approach was completely different from the conventional one that requires a student simply to learn the study material for a school test.

2 Pedagogical Background

This learning–teaching strategy is based on constructivist learning theory. The emphasis is on the process of students' active construction of knowledge. This means instigating a guided inquiry where students investigate teacher-presented questions, using student-designed procedures.

Our premise is that students are more often interested in studying when the study context and tasks are understood, realistic life situations are discussed, or relevant data are used. In this project students use Google Earth and Street View, and virtual globes with extending geoinformation services form a basis for spatial citizenship education. Satellite pictures, photos, and different perspectives, together with other information, give learners a quite naturalistic idea of remote places and their inhabitants. This task-based study encourages students to discuss how 'we here' could be involved with 'them there' through agriculture and trade. Students will find out, for example, how the Spanish grow cucumbers and tomatoes that we can purchase here in Estonia from a local shop. They are also required to discuss different environmental problems that are caused by particular production methods.

Studying is successful when the study environment is diverse, thus enabling a focus on different approaches and aspects. Subtasks given to students are neither too general nor too specific. This leaves an opportunity to investigate tasks separately and therefore to promote learning from each other. During the work process, the students have a chance to work individually at their own pace; on the other hand, they constantly need to work together and rely on each other's work. Each team member has his/her own assignment and thus a responsibility to the whole group. This study environment supports cooperation and communication rather than rivalry.

This study project is based on well-known principles: from local to global and from specific to general. Within the project, students work on different topics; from the daily purchase of foodstuff in Estonia, to a specific place in some remote country and aspects of agriculture; from foodstuff to agricultural production, environmental problems, and the role of agriculture in a country's economy. This approach eliminates possible limitations of conventional teaching, such as the formation of misconceptions and stereotypes – for example, that agriculture in developing countries is less developed and only self-sufficient.

The teacher's role in this study process is to guide and support, not to present knowledge. The teacher needs to encourage students and guide the development of their critical thinking, and to help students to develop their analytical and synthetic skills. During the teaching process, the clear aim is that the teacher also learns.

3 Description of the Learning Environment

All lessons took place in a computer class. Tasks, background materials and instructions were given using Moodle; the website was also used for uploading each group's completed work. Additionally, the Google Docs environment was used to exchange information between group members, enabling them to work together and to formalize their results.

In the introductory lesson, the students discussed the origin of the foodstuffs they personally, or their family, purchased daily from the shops. With the help of the teacher's guiding questions, students discussed why people buy so many different imported foods, why those imported vegetables and fruit are so much cheaper than those grown in Estonia, how production methods affect the environment, and why people could not give up agriculture. Students have an opportunity to debate where our daily food comes from: starting from the origin of milk and bananas, to smartphones that can be bought from shops, and the complex network of connectivity in a globalized world. To back up their thoughts and arguments, exemplary materials can be sourced from geomedia.

Next, the teams had an assignment to determine which countries import the most vegetables, fruit, dairy products and other foodstuffs. Students opened the homepage of Statistics Estonia (http://www.stat.ee/en) and used the foreign trade statistical database about the import of agricultural products to Estonia. Each team selected one product group and identified the countries that exported the most of those types of products to Estonia. They then prepared a thematic map showing the imports of different food products into Estonia. The teacher did not give very specific instructions, in order to enable the students to complete the task creatively.

During the next two lessons, the students worked with Google Earth in groups of 4. For this task, the teacher had prepared a Google Earth kmz-file (http://koolielu.ee/waramu/view/1-98c6b0f8-84a7-4ff8-b221-6b2e840c53c9) which had 30 marked places with different agricultural landscapes, for example olive oil plantations and greenhouses in Spain, feedlots and pivot irrigation fields in the USA (Fig. 6-1), terraced fields and rice fields in Vietnam, and so on. Two to three places per country were marked in order to stress the fact that agricultural production can be very diverse within a country. Each team selected two countries for research: one more economically developed (MEDC) and another less economically developed (LEDC); the second requirement was that both countries were located in different climate zones.

Figure 6-1: Zooming in and out on Google Earth enables one to see how the change in scale causes spatial generalization. Irrigation fields in Kansas, USA in various magnifications.

At the beginning, each team member was asked to describe the agricultural landscape in one marked place on the basis of the satellite picture. The teacher's instruction guided the students to describe the size and location of fields, built-up areas, infrastructure and other landscape elements. To obtain a better image of the place, pictures from Google Earth needed to be used or new ones had to be found on the Internet. Screen dumps from Street View could also be utilized (Fig. 6-2). With the help of the Internet or an atlas, the students then needed to determine what the main crops in the area were. The description of the landscape followed, and students decided on its farming type based on this information – e.g.

whether the area has subsistence or commercial agriculture, intensive or extensive agriculture. In addition, they tried to explain how the agricultural production in these places has resulted in the particular kinds of a landscape pattern.

Figure 6-2: StreetView enables students to associate map views in Google Earth with realistic images of agricultural landscape. Both can be depicted in their presentations. Greenhouse industry in Almeria, Spain.

Next, the students had to decide on the kind of effect that a farming practice has on the environment. Prior to this, they had examined the background materials prepared by the teacher, giving various overviews of agricultural effects on the environment. Each team had to decide which type of effect the place faces, and had to justify their decision. Team work results were presented in slides.

In the third stage, the students worked beyond the farming in a particular region, widening their study to compare farming at state level. Based on previous knowledge and the map, the students posed a hypothesis, i.e. which country has better natural conditions for farming, and in which country agriculture plays a substantial role in the economy. The students had to decide on which indicators to use in order to test their hypothesis. They could use hints in the teacher-compiled instructions, and links to FAOSTAT's agricultural resource bank (http://faostat.fao.org/default.aspx?lang=en) and the World Factbook. They needed to use various maps on the Internet and background information about the economy as well.

Based on thematic maps and climate charts, the students compared natural factors for developing farming in the two countries (relief, temperatures, precipitation and its distribution, the length of the growing season). Data on land use, agricultural production, yield and export of agricultural products were found from FAOSTAT. World Factbook was used to find data on farm workers, and the proportion of agriculture in

GDP was calculated. To give a better overview, diagrams had to be prepared (for example, for comparison of land use or yield). Following data collection, processing and analysis, the groups made slides of the comparison between agriculture in the two countries, and these were added to slides made in previous lessons.

In the final lessons, each group presented its work to their fellow classmates. In this way, the whole class received an overview of different farming types and their effects on the environment, and the role of agriculture in countries of different developmental stages. All students had an opportunity to ask questions and comment on the results. During the presentations, all students assessed the work of the other teams, using an evaluation guideline prepared by the teacher. Finally, an evaluation of their own work was also given. It was also a requirement that they recorded which skills, knowledge and attitudes this project developed.

4 Preparation for the Teacher

For the teacher, the preparation and teaching of the project is time-consuming because it requires significantly more than introducing the topic in a conventional way. This is one of the reasons why computer-based group work is not a very popular teaching method among some teachers.

The teacher has to prepare instructions in order to guide students to work independently with data, compare relevant factors, and make appropriate conclusions (http://koolielu.ee/waramu/view/1-d0a2575e-60a7-4193-ac2a-e23c1af95d76). Although some students can work independently, many students still need support, for example in constructing graphs or charts in Excel. Cooperation with teachers of mathematics and information technology could certainly help to increase the efficiency of these lessons.

Additionally, background materials that contain terms related to agriculture, and a systematic and diversified overview of agricultural effects on the environment have to be prepared. Students can find some help in textbooks, but the study material found there is often too general and prepared for other purposes. Unfortunately, Estonian teachers cannot use very many systematic background materials in their lessons, because the material is often unavailable in translation into Estonian. For example, we have been unable to find in Estonian, on the Internet, an overview of the farming types and environmental effects necessary for this project.

For the activities to work effectively, it is necessary for the selected agricultural landscapes on Google Earth to be easily understandable for students; the typical farming types must also meet the requirements of the list of farming types in the curriculum; also necessary is the availability of good-resolution satellite pictures and suitable photos of relevant places on Google Earth or on the Internet.

Most of the students did not have any experience of working with satellite pictures. They were therefore given instructions as to how to describe and identify agricultural landscapes from satellite images, and help in understanding which characteristics of a particular farming landscape needed to be studied in greater detail.

Guidelines for using the FAOSTAT resource bank were also prepared. Since many secondary students do not have fluent English, the teacher translated several indicators available in the database into Estonian in order to speed up the process. In general, more capable students can do this by themselves.

The instructions prepared for this study project could be easily modified and supplemented in the future, so this would significantly reduce the teachers' work in preparing similar lessons.

During the work, the teacher assessed each student's study process: how much work they could do during the lesson, the student's attitude, and how the work was divided between the group members. The results of the work done in lessons were uploaded to Moodle, where the teacher could make corrections and suggestions. At the end of the project, group work, reference to sources and presentation were all taken into account.

5 Conclusion: Experience and Proposals for the Future

This project has been used in Jaan Poska Gymnasium (Tartu, Estonia) several times in recent years. In 2010, a master's thesis was written about one part of the project – how secondary school students understand agricultural landscapes using Google Earth satellite pictures (Ilves, 2010). Over time, the structure of the project and its instructions have been improved, based on the results of the students' work and feedback analyses.

For students, the most attractive (and surprising) part of these lessons is the work with Google Earth (Fig. 6-3), for example, to explore modern commercial and intensive farming in less-developed countries.

As teachers, we observed students in computer classes. The research conducted by Ita Ilves (2010) showed that the interest in studying agricultural topics increased after examining agricultural landscapes on Google Earth. Many students noted that lessons with Google Earth had been exciting and had broadened their minds. The effect of Google Earth to pique students' interest and increase their learning motivation has also been noted in articles by Doering and Veletsianos (2007) and Patterson (2007).

Figure 6-3: Tea plantation and workers' dwellings on a satellite picture and a photo, Kenya.

Most of the students managed to describe the agricultural landscape shown in the satellite pictures; crops grown in these fields were also identified using the photos and the Internet. Students performed relatively well in identifying farming types (self-sufficient or commercial farming), and gave reasons for their answers. However, they had difficulties in finding argumentation on whether it was intensive or extensive farming.

To decide what effect the type of farming has on the regional environment, the students had to undertake prior reading of theoretical background materials prepared by the teacher. Many students found this task unachievable, because the knowledge obtained had to be implemented in a new situation. This is a skill that starts the process of independent thinking.

The students managed well with finding and downloading statistical materials. However, many faced difficulties with data processing, with making comparisons and with drawing appropriate conclusions. Absolute numbers were used frequently in comparisons; only a few teams drew charts and graphs to present the comparisons. The students also had difficulties with making correlations, finding arguments and drawing conclusions.

The students have under-developed skills of cooperation – too much time is spent on distributing tasks among team members. Feedback showed that they would prefer to work individually since all team members do not contribute to teamwork equally. In addition, the students are often unable to relate the results of co-workers to their own work.

This project, or elements of it, is certainly adaptable to different situations elsewhere. Studying and describing agricultural landscapes on

Google Earth is the most interesting part of the project, and probably the most valuable, for students. Implementing similar studies would be made more efficient if teachers could use kmz-files that included very typical farming types in different countries, along with photos of agricultural landscapes made especially for this study's purposes. Even though Google offers some possibilities for similar activities, the digital earth.eu project could be a good platform to create a depository of relevant educational resources, especially if undertaken together with teachers from different countries (http://sitescontent.google.com/google-earth-for-educators).

References

Doering, A. & Veletianos, G. (2007): 'An Investigation of the Use of Real-Time, Authentic Geospatial Data in the K-12 Classroom'. *Journal of Geography* 106 (6), 217–225.

Ilves, I. (2010): '"Formation of Students": Understanding of Different Agricultural Areas with the Support of Google Earth'. MSc dissertation: Tartu.

Patterson, T. C. (2007): 'Google Earth as (Not Just) a Geography Education Tool'. *Journal of Geography* 106 (4), 145–152.

CHAPTER SEVEN

APPLICATION OF VIRTUAL FIELD TRIPS IN TEACHER TRAINING AND GEOGRAPHY CLASSES

ALEXANDRA BUDKE

Abstract

Today, students as well as trainee teachers have easy access to digital media such as smartphones, iPads, laptops or Netbooks, which are used extensively to manage daily life. Having been brought up with digital media and employing them competently in many contexts, members of this generation are often called 'digital natives'. Klein (2008) observed an especially high interest among students in using the Internet and computers in geography classes. Educational institutions (schools, universities) have reacted to the virtual revolution by the revision of their curricula and the purchase of corresponding technology. However, there is still a deficit in the actual usage of the 'new media' in geography classes (Klein, 2008) and geography teacher training, due to a lack of didactical concepts and examples of how to deploy the media meaningfully in class. This is the starting point of the virtual field trip introduced in this paper. After an introduction of the concept, its usefulness in geography classes is discussed.

1 Pedagogical Background of Virtual Field Trips

Learning geographical subject matters through field trips has long been an important method in geography teacher training and geography classes. While traditionally the main goal was to learn geography by exposing students to 'real' spaces and landscapes, today field trips pursue the understanding of individual perceptions of space and socio-spatial constructs as a result of recent developments in geographical scientific paradigms (Ohl & Neeb, 2012; Budke & Wienecke, 2009). The application of virtual field trips is discussed in this context. They are characterized by offering the possibility to work interactively with spatial information on a given geographical problem. Usually, cartographic representations are used for navigation. Virtual field trips have a great potential for use in geography classes, because distant regions can be explored at little expense from the classroom, interesting and up-to-date material can be used, and students' abilities to learn in a self-directed way and interactively are enhanced (Budke & Kanwischer, 2006). Furthermore, their media competences can be strengthened through the evaluation of given materials (diagrams, maps, photos, films) and the production of self-made media, which is an important goal of geography classes. By the analysis of videos, texts and photos, the specific perspectives, interests and motives of different stakeholders become obvious. In this way, students are able to understand spatial conflicts and to evaluate them in a more sophisticated manner. They can learn to explain the real consequences of social and political constructions of space. An additional advantage is that students will only find tailor-made teaching materials in the virtual field trip and will not be overwhelmed by the vast number of different websites.

2 Description of the Learning Environment – A Virtual Field Trip to the River Rhine

The need for a suitable pedagogical concept arises in order to implement the benefits of virtual field trips. A few decisions of a theoretical nature made in the context of the production of a virtual field trip to the River Rhine follow. The project was sponsored by the Rhein-Energie Foundation and conducted jointly with the team for media informatics at the University for Applied Sciences, Cologne.

Conceptual Goal: First the topic of the field trip must be determined. In our case, this was the portrayal of the conflicting use of rivers typical for European cities. Our virtual field trip helps to experience this conflict, using Cologne as an example.

Objectives Concerning Competencies: Our field trip pursues the main pedagogical goal of strengthening students' argumentation competencies. Conflicting uses are outlined and the differing argumentations of the most important stakeholders are presented, because controversial perspectives are necessary for any argumentation. Analysing the arguments will improve the students' skills in understanding argument. In order to increase personal involvement, opportunities for debate among students are offered by inserting chat functionalities (competence in interactive argumentation). In this way, exchange with people outside the classroom will be made possible. This will confront the students with new points of view. Integrated tasks and exercise sheets further request the production of argumentation at each stage of the field trip, so competence in the production of argumentation will also be strengthened.

Selection of Perspectives. In a third step, a range of possibly conflicting points of view that are typical for our topic were identified, because conflicting uses occur whenever differing spatial perceptions and interests of the local stakeholders exist. In our case, these perspectives were (Figure 7-1):

- Ecology: From this perspective, the Rhine appears as an ecosystem to be conserved and used in a sustainable manner.
- Living: From this point of view, the Rhine is of interest mainly as a housing location and/or lucrative real estate investment. The question of how the river banks can be made even more attractive to its residents is raised.
- Flood risk: From this perspective, the Rhine poses possible threats to the residential population, which are to be minimized through flood protection measures.
- Recreation: The Rhine and its banks can be viewed as attractive recreational areas, offering space for leisure activities and tourism.
- Economy: From an economic point of view, the Rhine is one of the most important traffic routes in Europe. Cologne is the location of the second most important river port in Germany, strengthening the industrial location.

Selection of the Points of Interest. After having identified the typical social perspectives on the River Rhine, we selected locations which enabled us to understand the conflicts concerning the use of the river. We identified the port of Niehl, Worringer Bruch, the Old Town, Rheinauhafen and Rheinpark as suitable locations within the city of Cologne (Fig. 7-2).

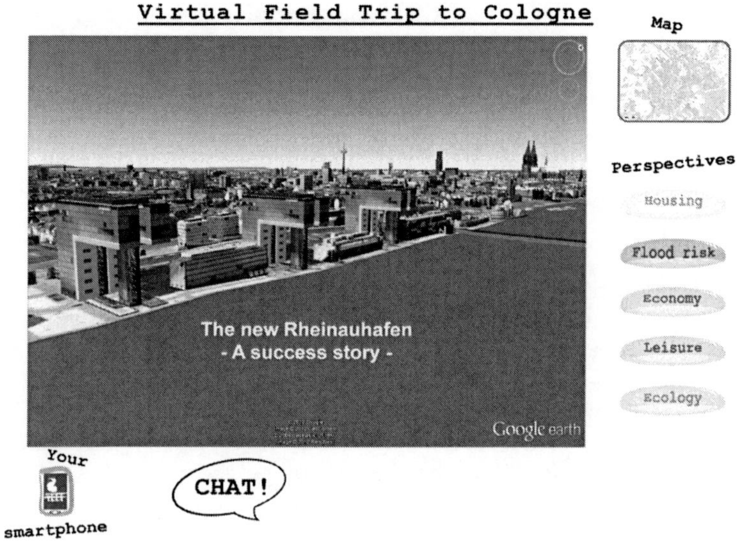

Figure 7-1: Perspectives concerning the ‘Rheinauhafen’ in Cologne (graphics: F.v.Reumont)

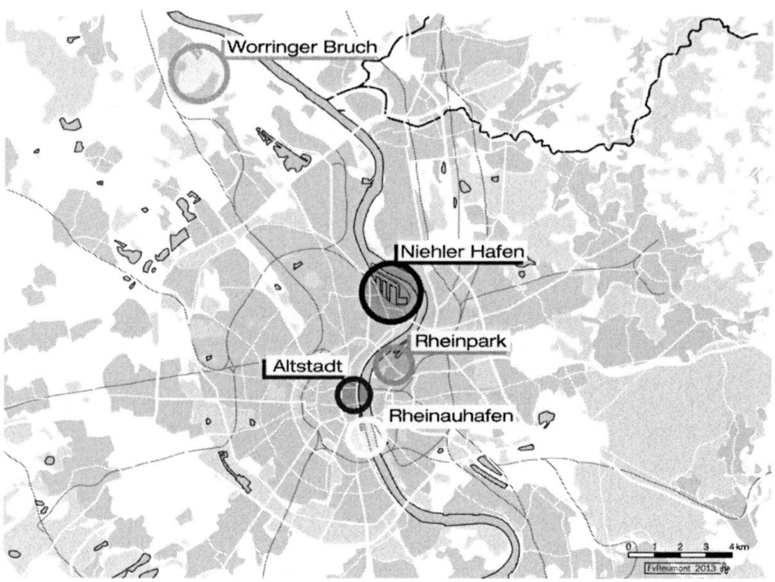

Figure 7-2: Points of interest for the virtual field trip to Cologne (graphics: F.v.Reumont)

The various perspectives with their respective lines of argumentation concerning a given problem are introduced at each stop of the virtual field trip. At the point of interest 'Port of Niehl', for example, the starting question is whether or not it makes sense to expand the river ports of Cologne. At this point the students could come to the conclusion, for example, that it does make sense from an economic point of view, but it makes less sense from an ecological, flood risk, recreational or residential point of view.

Technical Specifications. We chose the open source content management system Joomla as a platform for publishing the virtual field trip on the Internet. Joomla provides the technology to present the different media formats integrated into the field trip. Furthermore, Joomla is designed specifically for web2.0 applications, which is important for the installation of chat, wiki or forum functionality, allowing more opportunities for participation. We installed the system on a server, optimized it for multimedia content, and made it accessible to the media editors, so that they could upload their material directly using their personal accounts.

An essential constituent of the virtual field trip is cartographic representations emphasizing the manifold spatial relations. The basis for navigating through the virtual field trip is an interactive 2D map, which, slightly modified, is also used for the 3D city model. The data for the map originates from the OpenStreetMap project. We selected, generalized and designed it according to the needs of the field trip.

3D models of the city of Cologne and the field trip's different points of interest are the core components of the cartographic representation. The models illustrate, for example, the course of the river Rhine and show the position of Cologne in Europe. Additionally, the 3D models of each individual stop on the field trip are accessible to the participants, allowing them to roam freely and to explore them (Fig. 7-3).

The data of the city model were provided courtesy of the surveyor's office of the City of Cologne (Amt für Liegenschaften, Vermessung und Kataster). They were delivered in medium resolution (level of detail [LOD] 2) and without any textures. Buildings that are landmarks or of specific interest in the context of the field trip were later modelled in greater detail, enhanced with textures, and integrated into the model. The city model was placed on a digital elevation model (DEM), created from the freely available shuttle radar topography mission (SRTM) data. Alternatively, a high resolution DEM provided by the spatial data centre of the municipality could be used. An overlay of the DEM with texture representing a map of the field trip area is created. In this way a vivid and appealing 3D scene is created representing the area of interest.

Figure 7-3: 3D model showing the Old Town of Cologne (F.v. Reumont)

For the online real-time rendering of the 3D scenes, we reduced the resolution of the textures as well as the geometries that make up the city model, in order to save computational and transmission resources. The JavaScript framework x3dom (pronounced 'x-freedom', www.x3dom.org) was chosen to present the models online. It is based on the open ISO standard X3D mark-up language. The great advantage of this solution is the seamless integration of the 3D content into the virtual field trip's web app [since it is rendered directly in the browser] without the need for extra plug-ins. Another advantage is the integration of the 3D scene in the HTML tree structure, allowing for easy and efficient manipulation with JavaScript. x3dom makes use of the WebGL interface of the browser to deploy the client's 3D hardware acceleration. This facilitates a high performance representation of the 3D content. In cases where the browser does not support WebGL, a fall-back model using the Adobe Flash plug-in is provided.

3 Preparation for the Teacher

Just as with field trips into 'real' spaces, virtual field trips need thorough preparation and follow-up processing in class (see Figure 7-4). In the preparation phase, a problem-orientated introduction to the topic should be organized in order to activate the students' prior knowledge. For this purpose, a question that could be answered through work during the virtual field trip might be posed, e.g. 'What conflicts might arise from the exploitation of the Rhine?'. In order to prepare for the field trip, the students can then collect and share their knowledge about the typical users of the Rhine and conflicts among these users. In this way, the students learn more about the typical user groups associated with the Rhine and their interests (economic, ecological, leisure, flood protection, housing). They will then find out more about them in the course of the field trip.

Following the preparation phase, the students proceed to the actual field trip. The field trip can be worked on in class or at home. If the field trip is to be worked on in groups, these groups might be organized according to either the user groups of the Rhine or the different points of interest within the virtual field trip. Each student, or group of students, would then investigate either the interests of a specific user group (e.g. economy) at each point of interest, or the central conflict between the user groups at one of the points of interest. In order to secure results, for each point of interest quizzes and worksheets are provided. So that the students can communicate with each other and strengthen their argumentative skills while working on the virtual field trip, a 'chat function' was implemented that can be used at each point of interest.

The field trip can be found at: http://wasser-in-koeln.web.fh-koeln.de /joomla/index.php/de/.

In the post-research phase, the students try to answer the initial research question through correlating the information they have collected on different aims and interests of the user groups of the Rhine, thus making it possible to understand the different usage conflicts. For this purpose, different methods such as discussions, role plays, debates or concept maps can be used. Follow up questions could be: How did these conflicts emerge? What are the implications? How should the knowledge about the different interests be used for future planning?

As a conclusion, students and teachers should reflect on the learning process and the use of the virtual field trip.

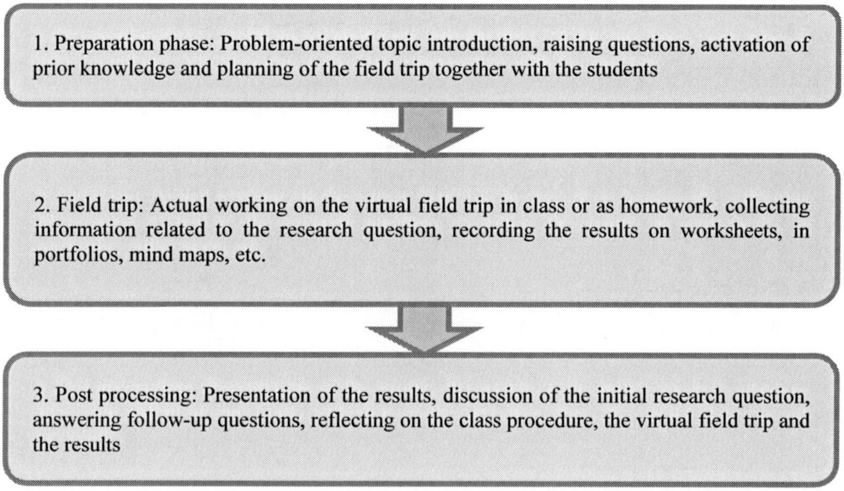

Figure 7-4: Overall structure of the Virtual Field Trip

5 Experiences and Alternative Tools

To date, the virtual Rhine field trip has not been evaluated. Evaluations of existing virtual field trips show that they can be used to deepen and consolidate knowledge acquired in class. Of course this is only possible if the content is appropriate, the materials are suitable for the group and the school type, and qualitative criteria such as problem-orientated procedures, controversy, interactivity, up-to-dateness and activity-based learning have been respected. The result of the evaluation in a ninth grade class (age 15) of a virtual field trip to Guadalajara (http://www.guido.uni-koeln.de/virtex/start.html) was that the students had found it mostly comprehensible and felt that it had deepened their knowledge about water-related problems in Mexico. Furthermore, they claimed that the field trip had awakened their interest in the topic. Additionally, they were exceptionally motivated to work on the virtual field trip.

As well as using existing virtual field trips, it is possible to let students create their own in group work. If publishing the virtual field trip on the Internet is not planned, this step can easily be replaced by using Power-Point software, which allows for various media and interactive elements to be integrated. However, publishing on the Internet poses few problems today because of the great variety of free software. In order to enable the students to use their own media and to respect copyright issues, a virtual field trip designed to be published on the Internet should be based on topics about the students' immediate environment. The concept allows a teacher to differentiate easily between performance capacity, working speed and personal interest of the students, as the number and complexity of the media to be integrated can be easily adapted to the needs of the individual student. A good overview of the technical possibilities in creating virtual field trips can be found in Chan and Dickmann (2010).

It is expected that students in primary and secondary education will work intensively on topic contents while working on a virtual field trip, and their media competence, in the sense of creating and critically reflecting on various media, will be reinforced. An appealing aspect of this work is that students become creatively involved in the process of producing new media which, after completion, can be made available for other learners.

6 Conclusion

Due to the speed of technical developments, the Internet nowadays offers an increasing number of interesting virtual field trips which, with appropriate preparation and follow-up processing, can be used profitably in geography classes. Furthermore, it is possible to include virtual field trips as appealing learning activities in geography teacher-training. This medium offers the chance for learners to acquire key content of geography education in an individual and self-determined way, as well as the possibility to exchange with people outside the classroom. In future research and development projects, the question of how virtual and real field trips can be usefully combined within the context of geography education will be an interesting issue.

References

Budke, A. (2009): 'Kompetenzentwicklung auf geographischen Exkursionen'. In A. Budke & M. Wienecke, *Exkursion selbst gemacht. Innovative Exkursionsmethoden für den Geographieunterricht*. Potsdam: Universitätsverlag.

Budke, A. & Kanwischer, D. (2006): '"Des Geographen Anfang und Ende ist und bleibt das Gelände!" Virtuelle Exkursionen contra reale Begegnungen'. In W. Hennings, D. Kanwischer & T. Rhode-Jüchtern, *Exkursionsdidaktik – innovativ!?*. Erweiterte Dokumentation zum HGD-Symposium 2005 in Bielefeld / Weingarten.

Chan, M. & Dickmann, F. (2010): *Virtuelle Exkursion im Internet. Die Nutzung frei zugänglicher Software für die Visualisierung räumlicher Informationen auf interaktiven Websiten.* Materialen zur Raumordnung, Bd. 73. Bochum: Universitätsverlag.

Klein, U. (2008): Geomedienkompetenz. Untersuchung zur Akzeptanz und Anwendung von Geomedien im Geographieunterricht unter besonderer Berücksichtigung moderner Informations- und Kommunikationstechniken. Kiel: Universitätsverlag.

Ohl, U. & Neeb, K. (2012): 'Exkursionsdidaktik: Medienvielfalt im Spektrum von Kognitivismus und Konstruktivismus'. In J. B. Haversath (ed.): *Geographiedidaktik*. Braunschweig: Westermann.

CHAPTER EIGHT

FLOODS: DEALING WITH A CONSTANT THREAT

ROLAND GOETZKE, HENRYK HODAM, ANDREAS RIENOW AND KERSTIN VOß

Abstract

Problems of vulnerability and risk are often analysed spatially using remote sensing data. Geomedia offer a wide range of methods for this purpose, but are not normally recognized in school education. The reasons for this are as follows: teachers are not taught how to use remote sensing data, nor are there material or software solutions which can be integrated into school lessons without additional preparation. There is a way out of this predicament, which is to create easy-to-use materials to apply remote sensing methods in schools. This is the main goal of the FIS (Fernerkundung in Schulen – Remote Sensing in Schools) project, a research project funded by the German Aerospace Centre (DLR) with financial support from the German Federal Ministry of Economics and Technology.

The project's general approach in creating learning materials is to use e-learning methods in order to support learning mechanisms based on moderate constructivist learning theory. The interactive learning unit 'Floods – Dealing with a Constant Threat' is a module for a double lesson that enables students aged 12 to 16 to use geo-data for a site analysis, focusing on the threat of floods. It incorporates different methods which are exemplary for the way different types of e-learning materials are created in FIS lessons

1 Aim, Target Standard, Target Group

High floods and their destructive consequences for local residents are broadcast periodically by the media and are, accordingly, noticed by students. This awareness is especially true of the discussions about the impact of man-made climate change on the frequency of extreme natural events, which reinforces the importance of addressing the subject in school curricula. On the one hand, floods are determined by regional and local natural conditions, but on the other hand they are influenced by the way people both use natural resources and deal with this constant threat. The trade-off between the economically inevitable use of land in a catchment and the need for protection against natural risks, and the implementation of environmental protection have made floods a topic in many curricula throughout Europe (e.g. Netherlands, Germany, UK).

Which areas are especially vulnerable? Which measures of flood protection can reduce the risk? Geomedia can help us to understand the spatial context and the human–environment relationships leading to extreme events like floods.

In this contribution, we present a computer-based learning module that enables students to strengthen their spatial thinking in a decision-making process. The aim of the teaching unit 'Floods – Dealing with a Constant Threat' is the evaluation of selected sites in the light of natural hazards. Using GIS methods and a digital elevation model within an interactive learning module, the computer is specifically used as tool and, thus, improves the students' media literacy. The teaching unit addresses students aged 12 to 16. The primary subject is Geography, although the learning material offers links to other subjects in the natural sciences (e.g. Physics or Ecology).

The students come into contact with the subject of floods in the context of complex real-world problems, as they play the role of experts who assist a company in finding new locations. Following the initial guiding task, the pedagogical background of this teaching unit is linked to Problem-Based Learning (PBL) theories and skills-orientated teaching. According to the principles of moderate constructivism (Karagiorgi & Symeou, 2005), the learning process of 'Floods' requires a high degree of self-organization and responsibility. The students are free to construct their own hypotheses based on spatial analysis, and the multi-media preparation of the learning materials supports their cognitive learning processes. The teaching unit makes use of intuitive user guidance, and provides different means of coding (texts, images, interactive animations, films, figures etc.). The module is designed in such a manner that at certain points the students are required to reflect on their decisions after incorporating new information. In this way, students use the added value of geo-data as well as GIS and remote sensing technology to answer a specific research question. They communicate their hypotheses via these technologies, using data as the basis for their argumentation and decision-making. Therefore, the teaching unit is useful not only to enhance (geo-)media skills, but also to improve communication, as well as to enhance judgment skills and spatial orientation. Thus, the use of geomedia goes well beyond being a mere illustrative teaching element in geography lessons.

The next section gives an overview of the learning environment and the general structure of the learning modules developed in the FIS project. In the third section, we share some of our experience with 'Floods – Dealing with a Constant Threat' and present feedback we received on the applicability of the learning modules in the classroom.

2 Description of the Learning Environment

General Structure. The learning module consists of four parts. Each one is separated from the next by a quiz. Tasks as well as background information direct each chapter to a clear aim and provide guidance for the learner. The first section introduces the main problem, which is to be solved in the course of the module. Central to the module is an 'analysis tool', similar to common software for spatial analysis. This tool was created to provide a simple user interface, along with the capability of being fully scalable to fit certain tasks. The image data as well as the functions are tailored to the requirements of the task, leaving out all unnecessary features of the tool. As the learner advances through the various stages, different sets of remote sensing data are analysed with different tools. The tools are easy to use and a tutorial offers short animations introducing each.

The pattern described is exemplary for most of the e-learning units created by the FIS project. It is the result of trying to incorporate ideas of moderate constructivism and of taking the findings of successful e-learning into account.

General Design Guidelines. By offering ways to openly interact with the material, including possibilities of making both right and wrong decisions, the learning process in the learning environment leads to self-construction of knowledge. This paradigm can be seen at work in the core of every section of the module, which is always centred on the analysis tool offering possibilities for self-guided analysis of geo-data. The actual work with geo-data takes place on a high level of interactivity, up to a simulation level.

The desirable degree of freedom for users exploring the possibilities of the environment is, however, hard to determine. According to De Jong

(2008), discovery learning has to be guided to be successful. As we made similar observations when working with interactive geomedia in classrooms, we decided to use several ways of scaffolding the learning process through directive support throughout the lessons. This scaffolding is implemented through a range of tasks, functional scaling, quizzes and additional background information.

Tasks. While they are free to test certain hypotheses, the pupils are given directions through tasks. These tasks suggest certain actions that the user should take using the analysis tools. They also give hints on how to use the available tools.

Functional scaling. Another way of directing progress in the learning environment is to limit the capabilities of the analysis tools. This decreases the extrinsic cognitive load by reducing operational features of the environment, and thereby limits the amount of possible combinations. It also prevents the learner from following hypotheses leading to dead ends.

Quizzes. Each section of the learning module concludes with a quiz which tests the facts learned so far. The learner is encouraged to match his or her recent experiences with new information. In this way, the knowledge acquired during the current section of the module can be consolidated and corrected. Thus, the quizzes are of high relevance to the learning process itself, because they ensure that the learner is prepared for the next part of the module.

Background Information. A high degree of interactivity is neither necessary nor appropriate for all types of content within the module. The content of the module that serves as background information employs animation accompanied by text. These parts of the lesson focus more on information acquisition than on knowledge construction. Therefore, a lower degree of interactivity is used (Clark & Mayer, 2008). Animations are not always the best way to convey new, complex material to the learner because they do not leave time to consolidate newly acquired information. To help fix the information in the student's mind, the animations are either segmented or looped in order to lower cognitive load (Moreno, 2007).

The Learning Module: 'Floods – Dealing with a Constant Threat'

Part 1. The initial screen shows a desk on which a letter and a mouse mat are placed (Figure 8-1). The letter contains the real-world task that the learner is to solve:

A confectionery manufacturer is looking for a suitable location to construct a new production site. The user is put in the position of an expert helping to find that very spot. Certain additional requirements are given so that the pupils have a starting point when beginning the general task.

After the pupils have familiarized themselves with the task, a click on the mouse leads them to the next section.

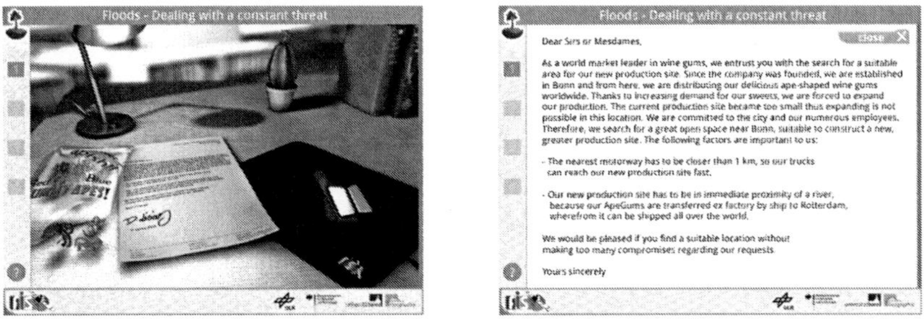

Figure 8-1: The initial screen (left) and the main task (right)

This part of the module serves as an introduction. Its purpose is to catch the learners' attention so that they immerse themselves in the situation and the task ahead. Presenting the general task graphically in a lifelike situation is meant to underline its authenticity, which meets the underlying pedagogical principles of PBL.

Part 2. The second part introduces the main tool. The geo-tool offers image data of the study area in the Rhine river basin in Germany. These image data consist of a satellite image taken by the Landsat-5 satellite and a thematic map of the area derived from this image by classification. The land-use classes can be identified by using an integrated tool. Two overlays show point and line information of possible construction sites and the location of motorways. A function that allows distances on the images to be measured completes the set of tools required to solve the task given in this part of the module.

The task is to decide which of the potential sites identified best fits the requirements; i.e. that the site for the plant should be close to a highway and well in reach of the river to provide easy transportation of the products.

The aim is to show how geoinformation can be used as an effective means to solve tasks based on spatial interrelations. Basically, the tool offers the users everything they need to solve the problem of choosing the new location of the plant. Although this location can, theoretically, be anywhere on the map, as the aim of the lesson is to introduce the threat of flooding, the possible spots from which the pupils can choose are limited to six, while one seems to be the perfect solution. This reduction in choice is necessary to guide the user towards the intended outcome of the learning module.

Part 3. Both images used in Part 2 are removed and replaced by a digital elevation model (DEM) and a map of the area showing only residential areas. These data can be easily compared using a newly introduced function called ‘swipe’ (Figure 8-2, left). Motorways and possible construction sites remain in the data pool. As measuring distances is not of relevance in this part of the learning module, this function is also removed.

Figure 8-2: Comparing images using the swipe function (left); background information on DEM generation (right)

As reading a DEM is most likely a new experience for pupils, an ‘Info’ section is available, conveying background information on how DEMs are produced and how they can be analysed (Figure 8-2, right). An interactive animation is used to make reading a DEM easier to understand. In this short film, an aeroplane scans a strip of terrain and produces a DEM. The possibility of looping the animation and the presence of a ‘pause-play’ button are user-friendly ways of helping learners to understand DEMs. They also give the learner more time to compare the terrain and the corresponding DEM.

The tasks in this part of the module are about using the DEM. Elevations are to be determined and the pupils are asked to make up their minds on how the elevation influences the location of built-up areas.

By introducing a new type of 3D geo-data, the initial tasks are extended by a third dimension. At first glance, this section seems unrelated because it does not deal with the basic problem; but understanding how to work with a DEM is a requirement for the incorporation of this data into the final decision-making. Going ahead with the task and including a new data type in a 'by-the-way' manner would have added unnecessarily to the cognitive load of the this part of the module.

Part 4. The DEM used in Part 3 remains in the data pool for further analysis, but now showing the water level instead of the ground level. The motorway and construction site layers are still in the data pool as well. The classified land use image and the distance measuring tool are reintroduced. A new analysis tool allows users to simulate flooding. This is done by marking values in the image data. Values within a certain range can be depicted as well, depending on the position of the 'similarity' slider. This gives the learner the opportunity to simulate flooding by marking the lowest elevation in the DEM situated inside the riverbed. Using the slider, areas that would be affected by flooding are marked (Figure 8-3, left).

Figure 8-3: Simulating flooding (left) in the analysis tool; interactive animation on flood risk (right)

The importance of taking the risk of flooding into account when planning new industrial or residential areas is explained in the 'Info' section (Fig. 8-3, right). Here, an interactive animation shows how man-made and natural risks of flooding are connected. Features like residential areas, rivers or dams can be removed or replaced in the model. Each action influences the general flood risk. This way, segmentation is implemented, and the learner can learn at his or her own pace.

The tasks lead back to the initial question, i.e. where to build the new factory. The pupils are asked to rethink their first choice, which took only spatial relations into account.

As the pupils are able to interpret a DEM by now, this newly attained knowledge can be put to work to solve the general task. They are asked to connect the flood risk to elevation and simulate a flood by using the data and tools provided. A new parameter is introduced that has to be connected to the knowledge acquired in the preceding sections. In Part 1, the pupils made use of geomedia in a straightforward way, measuring distances and using the information to reach a decision. Now they can use the data not only to make statements on current situations, but also to predict possible future developments. This broadens their view on the potential of geo-data. Above all, it shows how a problem cannot be approached and managed one-dimensionally, because the parameters to be considered are often multi-dimensional and complex. Now, faced with a new means at their disposal to predict the impact of a natural hazard, the pupils may be inclined to pick a different site for the new factory.

3 Preparation for the Teacher

The learning module is freely available via the website of the FIS project (www.fis.uni-bonn.de/EnglishMaterial). It is designed in Flash; thus, it is platform-independent, as it can be opened in any browser having a Flash plug-in installed. The downloadable zip-file contains the files needed to run the environment. For Windows PCs, there is a 'Floods.exe' file that can be executed directly. In this case, no Flash plug-in is needed. With other operating systems (MacOS, Linux), the file Floods.html can be opened in the web browser. The Flash plug-in is an absolute necessity in this case.

The average time required for conducting the teaching unit is 90 minutes. The lessons are accompanied by a didactic commentary and a set of sample solutions. Given the PBL nature of the module, the teacher's role is that of a facilitator: he or she observes the learning progress, assesses the pupils' individual learning behaviour, checks the pupils' intermediate results, and directs the discussion in the classroom. The teacher should be aware that some students may not used to working in a PBL environment and might not cope well with the 'confusion' and, hence, will need more time than others. Working in groups of up to four students at one computer can be beneficial for some students, while others might 'get lost'. For those, a small team of two students per computer is considered optimal. Nevertheless, experiences with this and other learning units developed in the FIS project have shown that the successful completion of the module in the classroom depends on good preparation and flexible implementation. With regard to preparation, the teacher should check the available computers in advance, especially whether the Flash plug-in is installed or the administrative rights are limited, to make the learning module accessible on every computer. Furthermore, the teacher should read the attached Teachers' Guide carefully and take a look at the sample solutions. This is especially important when students choose wrong solutions and come up with wrong results, which might leave them frustrated. Specific remote-sensing terms, like 'resolution' or 'infrared', can additionally be looked up in the FIS Info-Box, a digital encyclopaedia for the basics of remote sensing, which is accessible via the FIS Learning Portal (www.fis.uni-bonn.de). Because links to other subjects are obvious in the learning unit, it is valuable to discuss it with colleagues, especially from Physics and Biology. Ideally, they can contribute to the topic of floods in their subjects and teach topics related to meteorology, hydrology or ecology. For a better understanding of DEM-generation, the topic of laser-scanning or three-dimensional vision can also be covered in more detail in the Physics class.

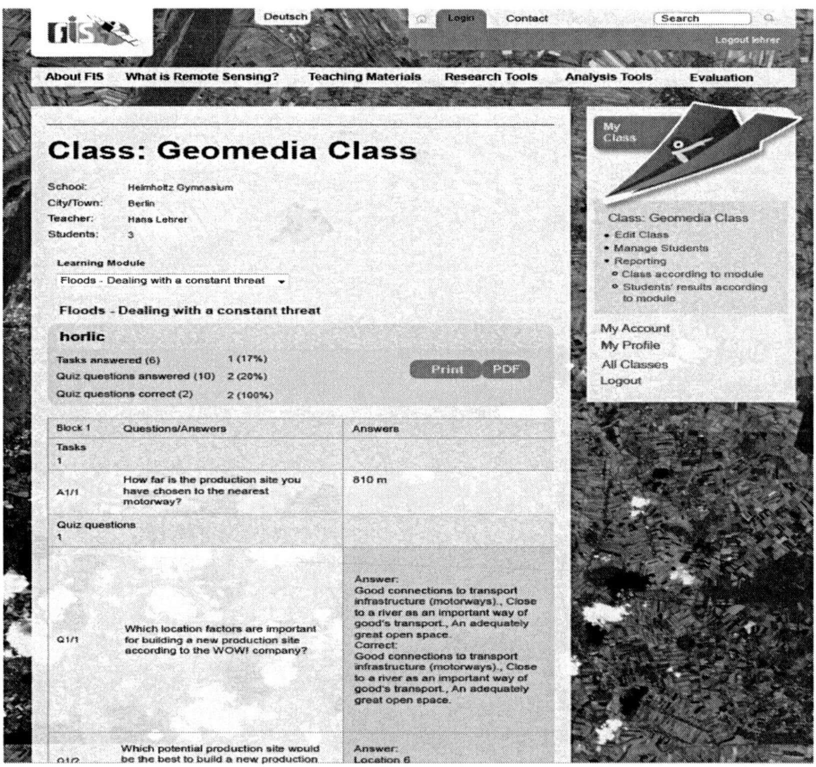

Figure 8-4: Screenshot of learning management functions in the FIS Learning Portal. Here: individual view of a student's answers in a learning module. Correct answers in a test are highlighted in green (here: light grey at the lower right).

Teachers in German-speaking countries can also make use of the capabilities of the FIS Learning Portal by navigating to the 'Geography' section and running the lesson completely online. Additionally, we implemented some basic learning management functions that are available after registration at the FIS Portal. For students, the advantage of registration is the automatic storage of the learning module's intermediate results. Usually, working with a learning module takes longer than one lesson. Registering with the learning portal allows users to save their work to date, to take up where they left off. When starting a learning module for the first time, a text box appears informing the user that intermediate results will be saved from now on. As well as the user's progress in the module, the quiz answers are also saved. While working on tasks, a pen symbol appears beneath the question, and the user is prompted to type in an answer. Having re-opened the module after stopping work, results and correct answers can be re-loaded into the module if desired. The benefit of registering for teachers is the opportunity to see, analyse and evaluate their students' intermediate and final results. For example, the teacher can assess the number of wrong answers for the quizzes, and read the pupils' answers to the tasks where the answers are not pre-defined (see Fig. 8-4).

4 Conclusion, Alternative Tools, Transferability

The aim of the FIS project is the creation of easy-to-use material for the use of remote-sensing methods in schools. The interactive lesson 'Floods – Dealing with a Constant Threat' shows how pedagogical theory, e-learning experience and technological possibilities can be combined to achieve that aim. The strategy applied to create the materials was established and tested in the early stages of the project. The learning unit presented here shows an alternative way to use geomedia, i.e. embedded in a ready-to-use environment that reduces the preparation time compared to the use of GIS or remote-sensing software. This way, teachers and pupils are able to use the benefits of remote-sensing software when working with geo-data, without being intimidated by complex computer programs. The learning unit shows how different kinds of media can be combined in an e-learning environment. It is centred on simulation-based interaction with geomedia, but texts, images and animations purposefully contribute to the learning experience as well.

Being a rather new lesson, the 'Floods' module still has to be evaluated with regard to its effectiveness. Surveys implemented on the online learning portal of the FIS project will shed light on this in the near future. First implementations at associated schools are already promising.

'Floods' is part of a larger portfolio of learning materials developed within the FIS project, all following the same teaching and learning principles and technical concepts. We receive regular feedback from collaborating teachers and their students in Germany regarding the applicability of the learning material. 'A new experience and a totally different way of working in the classroom' was one answer from a 12-year-old student in the online survey. For a project dealing with geomedia in school lessons, it is very important to know whether the satellite imagery used was comprehensible and suitable or not for the scope of the learning topic. Regarding this, nearly every person interviewed found that the satellite images were at least helpful for completing the learning unit. Astonishingly, few had conscious experience of satellite imagery before working with the remote-sensing data included in the FIS teaching units. However, we observed that many pupils were unaware that they were dealing with satellite data when using GoogleEarth. The teachers involved in the evaluation process found the FIS materials such 'Floods' relatively easy to use, where other software solutions with a focus on geo-information often lack relevance when integrated into everyday teaching. Similar studies have shown that existing software packages are often not intuitive and can be intimidating for people inexperienced with digital image processing who lack additional background information (National Research Council, 2006). Furthermore, many teachers have to struggle with organizational problems like administration permissions, diverging operating systems, or long-term timetabling.

Nevertheless, it is criticism which stimulates educational projects. We received very helpful suggestions related to different topics. For example, the use of very open question types, which are more appropriate at final examinations, was recommended instead of simple questions like 'what', 'where' or 'when'. Additionally, it would be better to avoid a strict definition of the potential audience – i.e. to adapt the materials for a whole range of educational levels, rather than restrict them to a particular class level. The duration of a double lesson is viewed critically as well, and a preference for shorter sections mentioned. Other suggestions related to technical aspects, like the speed of the video tutorials. Although an accepted paradigm in today's school lessons, some teachers struggled with the PBL approach to some extent. They noted a preference for clearly graded tasks, facilitating the pupil's assessment. Thus, it is a constitutive element of FIS not only to close the gap between knowledge and application of geo-data but also to mediate between the theory and praxis of learning conceptions in the digital age.

References

Clark, R. C. & Mayer, R. E. (2008): *E-Learning and the Science of Instruction: Proven Guidelines for Consumers and Designers of Multimedia Learning.* San Francisco, CA: Pfeiffer.

De Jong, T. (2005): *The Guided Discovery Principle in Multimedia Learning: Introduction to Multimedia Learning.* Cambridge: Cambridge University Press.

Karagiorgi, Y. & Symeou, L. (2005): 'Translating Constructivism into Instructional Design: Potential and Limitations.' *Educational Technology & Society* 8 (1), 17–27.

Moreno, R. (2007): 'Optimising Learning from Animation by Minimising Cognitive Load: Cognitive and Affective Consequences of Signalling and Segmentation Methods.' *Applied Cognitive Psychology* 21, 765–781.

National Research Council (2006): *Learning to Think Spatially: GIS as a Support System in the K-12 Curriculum.* Washington, DC: National Academies Press.

CHAPTER NINE

USING A PARTICIPATORY MODELLING APPROACH WITH GIS: WHICH PERSPECTIVES FOR SUSTAINABLE DEVELOPMENT EDUCATION?

SYLVAIN GENEVOIS AND CYRILLE CHOPIN

Abstract

This article presents the results of an experiment that we carried out in a secondary school in France in order to introduce a participatory modelling approach from a sustainable development education perspective. The case study was conducted using a free Geographic Information System (Quantum GIS) and concerned a planning project: the creation of a nature reserve in Brittany. Students of two classes in lower secondary education were involved in a problem-based situation and were invited to build their spatial visions using the GIS. As municipal representatives, they had to choose whether they wanted to be involved in this future reserve or not. The results of this educational research show that participatory modelling approaches are very useful in developing 'democratic' skills like participation and negotiation and can be adapted to other case studies to develop education for choice.

In France, as in many other countries, the use of geomedia in schools is largely dominated by the visualization of geospatial data using virtual globes or national geoportals. The use of GISs in secondary education is much less developed, even though they can integrate modelling and systemic approaches. GISs are most suitable for problem- or inquiry-based learning. However, cartographic uses in secondary education are mainly orientated towards descriptive or inductive approaches. So teachers have to overcome ritualized mapping practices in schools, in order to promote a new type of geography education (Genevois, 2008).

The educational experiment that we carried out is of potential use for teachers and learners who are looking for a reflexive use of GIS, which can be powerful tools for helping decision-making. We will present our case study from the participatory modelling approach first of all, and then from the point of view of educational design and the choice of tools and data. Finally we will focus on the results and the transferability of this empirical research.

1 Why Integrate a Participatory Modelling Approach in a GIS?

> Participatory modelling is a method that encourages the involvement of a group in the development of a model in order to improve understanding of a particular system, of its problems and its possible solutions, which will directly or indirectly lead to better management decisions. (HarmoniCOP, 2003)

The ARDI method (Stakeholders – Resources – Dynamics – Interactions), a participatory modelling method built by a research network in France, consists of three steps (Etienne, 2009). In the first step, users explore the data in the GIS and identify the territorial stakes at different scales. In the second step, they produce diagrams for discussion and discuss their spatial representations. In the third step, users have to arrive at a global understanding of the problems and find points of consensus among them. This method has been used primarily with stakeholders in fieldwork and with professionals for solving usage conflicts. It seemed interesting to transpose this participatory modelling protocol to the school context in order to lead students to consider the territory as a system, prior to making decisions and formulating proposals. Adaptation to a school context mainly consisted in helping students in establishing a spatial analysis. The challenge was also to use the GIS as a tool for producing and sharing spatial representations.

As underlined by Etienne (2009), the ARDI method is based on the idea that it is possible to reduce the tensions between stakeholders, if they have enough time for discussion. Collective construction of diagrams is intended to promote group dynamics. If the objective is to facilitate exchanges and interactions between participants, the implementation of a role play may be interesting. In our case, we preferred not to lock students into a role that would have led to partial and distorted views of land use. Nevertheless, we wanted the students to be put in a position where they would be able to make choices and see their consequences: education for choice is fundamental in sustainable development education. When the choice is effective, participants must be able to access information and to create knowledge themselves.

2 Design of the Case Study

The teaching–learning sequence that we built was conducted with two school classes of 30 students in the last year of lower secondary education in France (14–15 years old). This sequence is based on a real situation. In 2008, the Brittany Region initiated a project for a *parc régional naturel* (a regional nature reserve) whose scope includes 66 municipalities located along the lower course of the Rance river and the côte d'Emeraude, the coastline around the estuary of the river. Although a majority of municipalities accepted to join the project, some refused to be included within the boundaries of the future reserve. It was this deadlock situation that was our starting point for studying the stakeholders' conflicts relating to the protection of natural and cultural heritage in the context of sustainable development.

Specifically, the sequence was broken down into three steps. The first stage leads students to take ownership of the project's territory. For this, they work independently in groups of three and are guided to write a description of the territory. This phase of work is done in two hours. The teacher's role is limited to that of an advisor. In the second stage (also in two hours), students have to create a diagram of the interactions of various actors in the territory based on the use of the area, namely its natural and cultural heritage. Their task is to determine collectively who the local actors are and to separate direct and indirect stakeholders so as to determine their respective weights in the decision process. The area concerned by the project and the dynamics that contribute to transform it are described. This second stage, conducted by the teacher, allows the work carried out independently by students in the first step to be shared. The interaction diagram, with its system of relations, is intended to give a visual representation of the final project. Its development is gradual and more or less complete, depending on the levels of the students and their ability to detect multiple and complex relationships between actors. Formalization is provided by the teacher, who simply plays the role of mediator between different groups so that the systematic representation developed by the students is actually shared by them and not imposed. The goal here is clearly not to introduce an external element that would influence the next step. The third step is to place the students in the position of decision makers. It is for them to determine whether barriers to the finalization of the regional park project are related to defining the perimeter of the latter (and hence to its territorial dimension), or are simply due to insufficient cooperation between the actors they have identified. Students must make proposals to remedy the dysfunctions or reduce stakeholders' conflicts.

3 Choice of the GIS Software and Data

We chose to work with the geographic information system Quantum GIS (Q-GIS). This is free software commonly used in school. Q-GIS provides a good compromise between scientific requirements (the possibility of making maps with discretizations and spatial queries) and school requirements for a simple and fairly intuitive interface, available in different languages. Data made available to students is provided by different national institutes in France (National Geographic Institute, Institut national de la statistique et des études économiques [INSEE] etc.) or European agencies (data from the Corine Land Cover programme). Some data needed simply to be reformatted and renamed in order to be understood by students. Other data required more detailed work, like information layers obtained by joining tables, or the design of particular maps with diagrams.

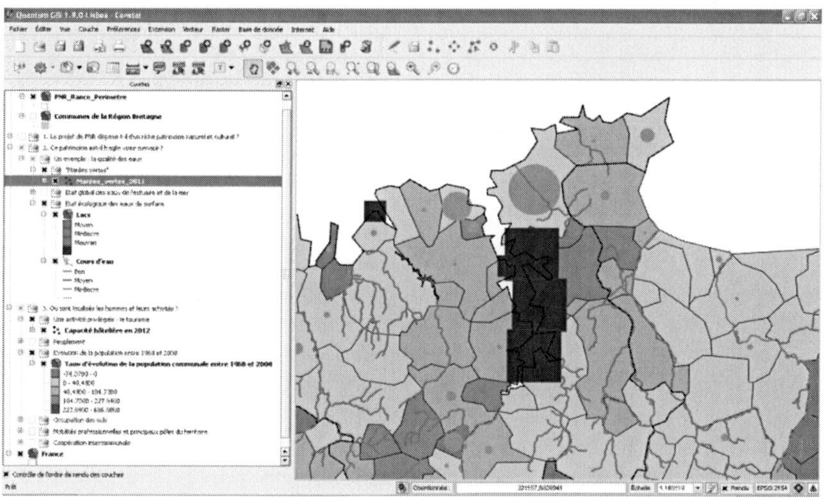

Figure 9-1: Simplified interface of Q-GIS and geographic information layers

GIS was used in two ways by students (Figure 9-2). In the first stage of their work, they had to describe the territory of the regional park using only data that were supplied. GIS was at this point a tool for understanding and above all for the appropriation of space. In the second stage, especially during the building of a diagram showing the inter-relatedness of the issues being examined, it also became a diagnostic tool for analysing stakeholders' conflicts.

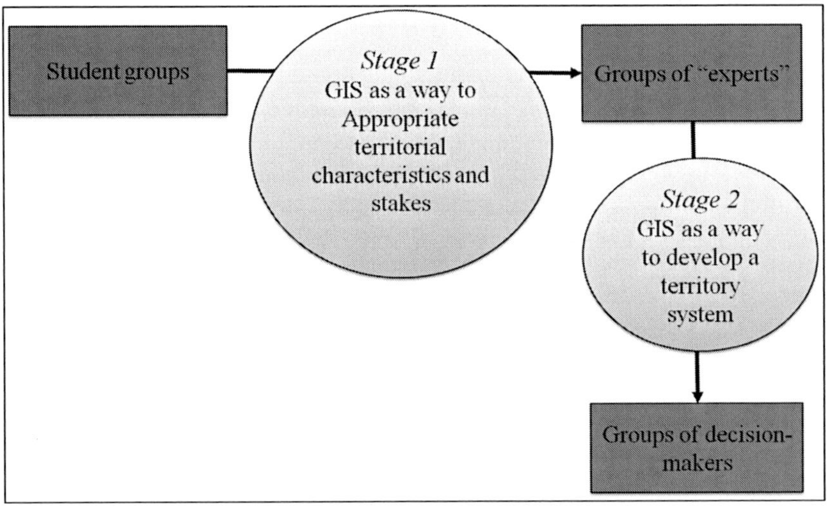

Figure 9-2: GIS as a way to change roles.

These two stages correspond to two distinct uses of GIS. During the first phase, the approach is essentially inductive. A significant number of information layers are subjected to the students' analysis. It seems difficult to offer students small data sets without jeopardizing their global understanding of land use. Without it, it is impossible for students to properly consider the deadlock situation and to propose ways to resolve it. To overcome the difficulties related to the number of layers, we grouped them into folders with titles, formulated as questions, so that students could be guided in their analysis of space and thus in its appropriation by them. The support by the teacher is facilitated by the explicit questions implemented in the geographic information layers: his/her role is essentially limited to explaining why the information provided by a particular layer would provide some answers to the questions. The layer which allowed the identification of bogs within Brittany caused great confusion among students. It was resolved by the teacher by making the link between this particular soil formation and biodiversity. It also at times fell to the teacher to correlate quantitative and qualitative information. Thus, for instance, some groups of students did not grasp that the number of hotel beds could be used as an indicator of the importance of tourism in the municipalities of the study area. However, despite the large number of information layers, students were able to appropriate the data sets, as shown by their capacity to apply them properly by themselves during the second phase of work.

In the second phase, the approach was different, because it required going back and forth between the model currently in development and the data provided. For example, the collective analysis of several layers such as agricultural activities and quality of surface water, allowed them to establish a causal link between land use and water pollution, and especially to highlight the inter-relationship between certain stakeholders, including farmers and tourism professionals. Correlated with urban sprawl, tourist pressure was therefore a challenge for the stakeholders who had to balance human activities and environmental protection. In using GIS, students were able to display both qualitative and quantitative data and compare them in order to find answers to their questions. Other geomedia, like virtual globes, are essentially data visualization tools and orientated towards qualitative information; they do not therefore offer the same flexibility in processing and correlating geographic information. The use of a GIS of professional quality is also required if we want to make students think about the semiology and choice of types of map graphics.

4 Results of the Experiment

It is always difficult to measure the contribution of the use of a particular tool, here the use of GIS software, in the success of an educational approach. The first two steps of the approach that we adopted aimed to elaborate collectively a systematic interpretation of the territory. The next step was to use the spatial model developed as a tool for decision support. Without offering a definitive answer, it seemed possible to make a first evaluation of this approach and of the use of GIS through the examination of the students' work. First, the synthesis of the different interaction diagrams made collectively by the students demonstrates great complexity and relative completeness (see Fig. 9-3). It therefore seems possible to consider GIS an effective tool to make a diagnosis of the land use, a perception that is strengthened by the students' comments. Indeed, in two classrooms, the same interpretation prevailed. The deadlock observed in the actual process of implementing the regional park is not related to the issue of the boundaries of the future park, but in the usage and stakeholders' conflicts which exist in this area. That is why the students developed proposals to reduce these conflicts. Thus, in order to preserve in a more sustainable way the natural and cultural heritage of the future park,

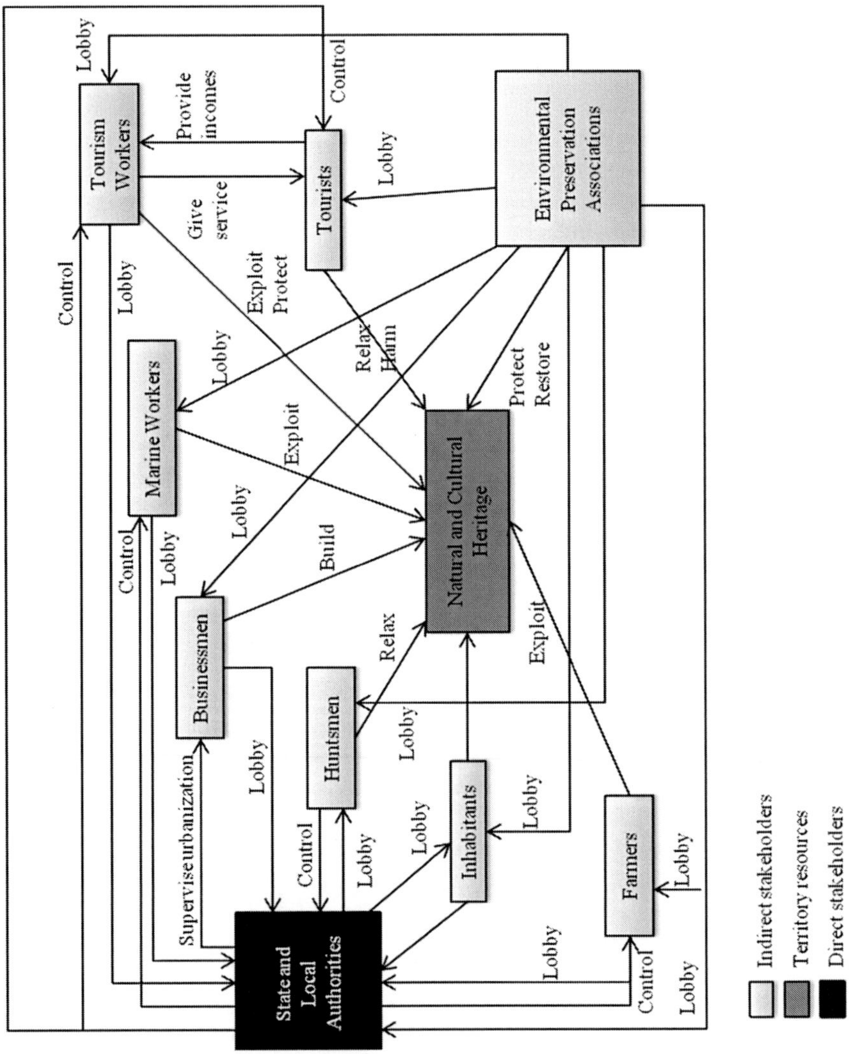

Figure 9-3: Synthesis of the interaction diagrams.

one of the two classes devised a system of financial equalization between tourism professionals and farmers to finance changes in agricultural practices in order to preserve coastal waters. Awareness of the usage conflict and the evidence for it at several different levels revealed by the GIS are at the origin of such a proposal. In the other class, taking into account the model developed collectively, some groups of students discussed a proposal from one of them. Indeed, in this case one group proposed a complete sanctuarization of the already protected areas, a proposal which the other groups of students immediately opposed, because their priority was to reconcile the interests of the different stakeholders in this area.

5 Transferability of the Case Study

The case study we carried out is transferable to other curricula, with other tools and other approaches.

The sequence described above implements a local case study adapted to the French Geography syllabus for students in *troisième* (14–15 years old). This syllabus contains an introductory case study of a regional or national *parc naturel* (nature reserve or national park). *Parcs naturels* reflect an urban vision of rural areas; the urban vision projects on to these parks its own concerns and urban practices. The sequence associates broader sustainable development issues, which are widely promoted in the secondary school curriculum in France in Geography and in other disciplines, such as Life and Earth Sciences. This approach is also widespread in European curricula (Krause, 2010). The sequence therefore addresses issues such as stakeholder or usage conflicts from a perspective of sustainable development in a context where inter-territoriality is particularly significant (Vanier, 2007). It thus brings to the fore the issue of sustainable development at the level of territorial cooperation between municipal territories. These issues are shared at different levels, particularly in Europe, and cover various issues that are not limited to a geographic approach. Other disciplines related to environmental studies or civics are also concerned at national or European level. They can therefore consider a similar approach using geographic information systems.

Other choices could be made by teachers who wanted to experiment with this case study. The systematic model developed by the students could be verified or modified by fieldwork in order to validate (or not) the hypotheses built using the GIS. In the ARDI method (Etienne, 2009), the use of a 3D model of the field is suggested in order to give a realistic representation of it. One can also use multi-agent systems (MSA) for analysing in more detail the impact on the landscape of selected scenarios. Simulation software could be used, too, although these types of tools might be considered too complicated for secondary education students.

References

Etienne, M. (2009): *Co-construction d'un modèle d'accompagnement selon la méthode ARDI: guide méthodologique*. Avignon: ComMod, INRA.

Etienne, M. (2010): *Companion Modelling. A Participatory Approach Supporting Sustainable Development.* Versailles: Quae.

Genevois, S. (2008): 'Quand la géomatique rentre en classe. Usages cartographiques et nouvelle éducation géographique dans l'enseignement secondaire'. Doctoral thesis, University of Saint-Etienne.

HarmoniCOP (2003): *Learning Together to Manage Together – Improving Participation in Water Management*. Osnabrück University.

Krause, J. (2010): *European Development Education Monitoring Report*, DEEEP; www.coe.int/t/dg4/nscentre/ge/DE_Watch.pdf

Vanier, M. (2007): *Le Pouvoir des territoires. Essai sur l'interterritorialité.* Paris: Economica.

CHAPTER TEN

EFFECTS OF TRAFFIC-RELATED NOISE: A GIS-SUPPORTED, PROBLEM-ORIENTATED SPATIAL ANALYSIS IN THE LOCAL AREA

JULIAN BETTE, STEFFEN HÖHNLE AND JAN CHRISTOPH SCHUBERT

Abstract

This article is concerned with the analysis of something that is both an important environmental problem and a local issue in the school area: traffic-related noise. It examines the particular opportunities which a full-function GIS delivers for analysing this space. Initially, noise measurements were taken in the area around the school. The data collected was then spatially analysed using GIS with regard to coincidences of traffic congestion. The teaching method followed Anderson's approach of inquiry-based learning (Anderson, 2007), which starts with a problem or a question and hypotheses related to the question. The systematic examination of these hypotheses provided a structure for the lessons. The objective data and the seeming spatial coincidences we discovered were questioned critically in the last step by contrasting them with subjective comments on exposure to noise.

1 The Teaching Unit: Introduction and Goals

Noise exposure caused by road traffic is considered a significant environmental problem EU-wide – a problem which not only leads to health-related damage, but additionally causes enormous economic and social costs (European Commission, 2011). Against this background, a common European noise mitigation policy (see Directive 2002/49/EC) has been introduced which binds the member states to, among other things, inform the public about noise and the effects of noise exposure, and conduct noise-mapping recordings on main roads, on a regular basis, in order to be able to identify the need for counter-measures. This process is accompanied by extensive EU-wide spatial monitoring in which every citizen can participate by importing his or her own primary data (see http://noise.eionet.europa.eu/ for example). The up-to-dateness as well as the political and societal relevance offer the potential to explore the topic of traffic-related noise exposure with students. This includes not only the personal attitudes of the students to the topic, but also the possibility of making an abstract EU guideline more concrete, through personal inquiries and analyses. We would argue that there is enormous potential in this respect. Geomedia are used in the inquiry (noise measurement using a smartphone app, geo-referencing using the smartphone's GPS) as well as in various steps of the analysis and the presentation of the results (GIS software). The technical applications used have the supporting role of tools throughout the teaching unit.

Our content-related goal is the analysis of the students' urban environment and the effects of noise exposure, based on the data the students have compiled. While undertaking the activity, the students analyse the relations between human beings (individuals, social groups) and the (physical) environment in the framework of a 'human–environment system'. Noise is caused by people in different ways, has spatially different effects on people, and is perceived differently by different people. The spatial effects of the distribution of the environmental factor noise on the distribution of social factors like rent levels or the quality of building structure are examined first by analysing spatial correlations between independent factors and dependent factors. The following deepening qualitative examination shines a brighter light on these relationships and helps to analyse whether these relationships are really relevant. Local transferability is ensured, and the integration of the European noise mitigation policy into the activity is made possible. The main target group for the teaching module is senior classes, because of the extent and depth of the analyses. The amount of time spent on the inquiry can be variable, depending on the complexity of the questions chosen; however, project work is the suggested approach. In addition to subject or content aspects, and with respect to analysing noise and noise exposure, students learn to work cooperatively and at a social level on a collaborative project. At the methodological level, subject-specific methods and procedures (noise measurements, GIS-supported spatial analyses, surveys …) are to the fore. The whole approach is characterized and integrated by a project-orientated structure and student-centred work. Thus, the students plan and organize the course of action autonomously, starting from their own questions and hypotheses, and continue through to the preparation and presentation of the results. In this competency-orientated geography teaching, the ability to solve problems independently is particularly important. By exploring different perspectives on space in the last step of the unit, the hypotheses-testing approach and the objectivity of data are questioned critically, thus significantly promoting metacognition.

2 Description and Illustration of the Teaching Unit

The following remarks illustrate the basic steps of problem-centred spatial analysis within a framework of project-orientated teaching. These steps can be modified or shortened at different points, depending on the learning groups, the questions examined, etc. The teaching process is structured in the following stages: (1) identification/extraction of the problem; (2) analysis of the problem and generation of hypotheses; (3) development and discussion of proposed solutions; (4) practical execution of proposed solutions; (5) closer examination; (6) generalization; (7) review and reflection on course of action (see Figure 10-1).

(1) The initial phase aims at identifying a problem-orientated question and, for motivational reasons, should have connections to the prior knowledge and interests of the students in a way closely related to their everyday life. Starting from a spatial phenomenon which is presented to the students by, for example, confronting them with newspaper headlines about traffic-originated noise exposure in their home town, a problem-centred question can be developed which refers to the effects of noise exposure. Attention should be paid to how the question itself is formulated: it should allow the formulation of testable hypotheses. It is important that the question should not only refer to the spatial manifestation of single factors (e.g. noise exposure in the city area) but also extend beyond this descriptive level and include spatial causes and effects.

Figure 10-1 (see over): Course of action in a GIS-supported spatial analysis in the local area (adapted from Bette et al., 2013)

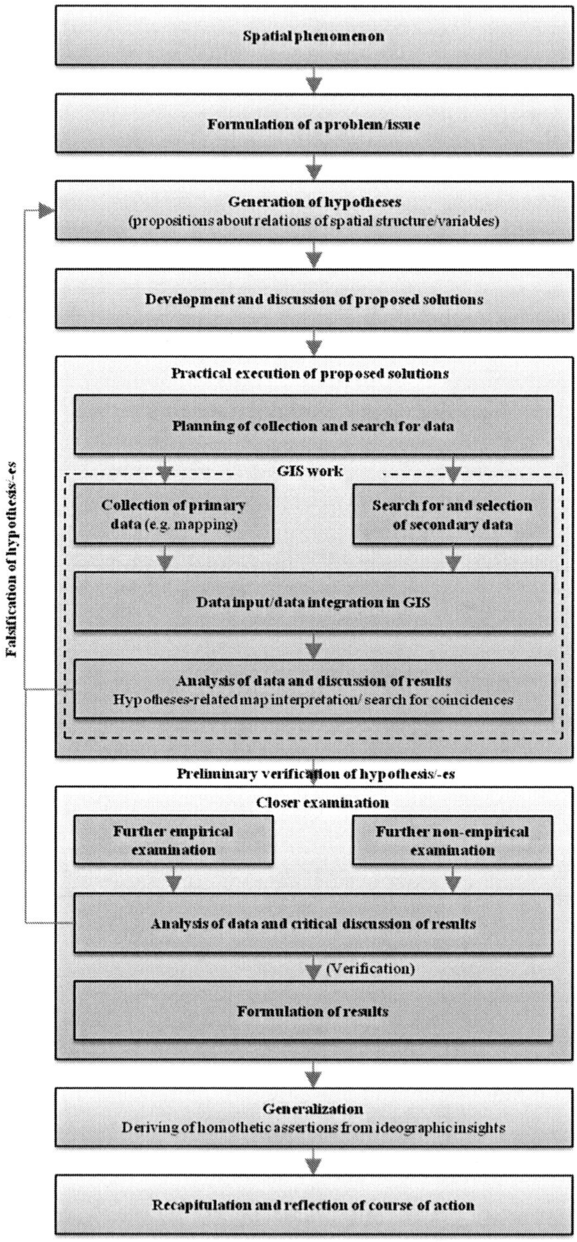
Spatial phenomenon
Formulation of a problem/issue
Generation of hypotheses
(propositions about relations of spatial structure/variables)
Development and discussion of proposed solutions
Practical execution of proposed solutions
Planning of collection and search for data
GIS work
Collection of primary data (e.g. mapping)
Search for and selection of secondary data
Data input/data integration in GIS
Analysis of data and discussion of results
Hypotheses-related map interpretation/ search for coincidences
Falsification of hypothesis/-es
Preliminary verification of hypothesis/-es
Closer examination
Further empirical examination
Further non-empirical examination
Analysis of data and critical discussion of results
(Verification)
Formulation of results
Generalization
Deriving of homothetic assertions from ideographic insights
Recapitulation and reflection of course of action

(2) At the outset, the problem of noise exposure is explored, based on the students' prior knowledge. If this prior knowledge is limited, the teacher can provide further material, e.g. a complete newspaper article rather than a headline about traffic congestion. Together the students discuss which factors determine noise exposure (e.g. number / type of vehicle, time of day, distance from road, etc.) and identify the effects that noise exposure has (e.g. rental prices, vacancy rate, quality of building structure, health implications, etc.). On this basis, the students can develop their own hypotheses about the spatial effects of noise exposure, which are operationalized and concretized in class (e.g. building structure is of lower quality in areas of high noise exposure than in areas of low noise exposure). The hypotheses should be well reasoned and thus not be formulated randomly; the focus in this step is, rather, on a proposition about the existence of spatial coincidences which suggest substantial relationships. Underlying complex contexts of justification will not be taken into account explicitly until step 5. During the formulation of hypotheses, particular attention should be paid to ensure that the hypotheses are testable or measurable, using collected or researched data and appropriate analyses. In doing so, space is deconstructed into relevant, measurable spatial factors (e.g. noise intensity, quality of building structure) or spatial structures respectively, in terms of an analytical-deconstructive approach (see Fig. 10-2).

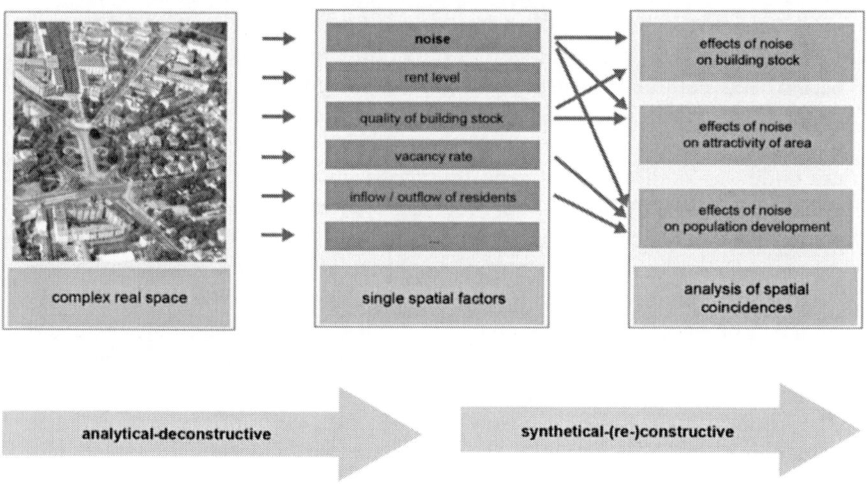

Figure 10-2: Schematic depiction of the approach in an analysis of coincidences/testing a hypothesis. Independent factor printed in bold. (Adapted from Siegmund/Naumann 2009, p. 8, Photo: H. Mühlpfordt 2001; CC BY-SA 2.0)

(3) The theoretical walk-through of proposed solutions by the students is the next step. For this purpose, it is helpful if the students can evaluate basic characteristics of desktop GIS applications and thus their usage potential. In an ideal situation, this will have been addressed in previous lessons. First, the proposed solutions are discussed, and pros and cons are balanced. In the process, the students discuss, in a well-founded way, which data are needed for the testing of the hypotheses and which area it would be appropriate to study. It makes sense to select an area with varying degrees of noise (e.g. main road and its traffic-calmed side roads). Comparing specific areas with high and low noise levels, and having regard for possible dependent factors (e.g. building stock as an indicator for income), makes it easier to suggest the possible effects of noise. The selection of measuring instruments or methods (e.g. measurement of noise level using smartphones, traffic counts, survey, observations, etc.) and the precise approach are planned and, of course, discussed. At this point it is also necessary to reflect on when a hypothesis can be considered (preliminarily) verified. This will usually be the case if spatial coincidences of structures (points, lines, areas) can be found.

(4) The next step is the execution of the proposed solution(s). There are two principal ways of doing this; both of them are typically used.

(a) The necessary geo-data for the solution (e.g. geo-referenced noise measurement using a smartphone app) are collected by the student and are entered into a GIS application (e.g. Quantum-GIS, GDV Spatial Commander). This practical step of data collection constitutes an essential part of the complete project, but can be assigned as a homework task, depending on the time available and the extent of the data collection. It makes sense to conduct the measurements using smartphones, since they are widely owned and used, and there is a large selection of appropriate apps which are available free of charge. From a political-participatory standpoint, the app Noise Meter, published by the European Environment Agency (EEA), is one we recommend. There is a facility for the collected data to be uploaded to a collaborative database or website of the EEA. This can be accessed by scientists and other users at a later date. The app is also designed to be very intuitive. The measurements of noise and locations for recording should be distributed to different student groups. The inquiry into noise exposure on site not only offers motivational advantages, but also connects th e objective data (sound pressure level / noise in dB) and the resulting map (see following steps) to individual impressions of the study area. This makes possible a critical view on the technical approach and the visualization of real space in the form of a map, because a comparison between data or map and one's own impressions of space can be made.

(b) If it is necessary, or seems to be meaningful, secondary data (e.g. from official sources) can be gathered and integrated in the GIS together with the student's own collected data. An appropriate base for the map can be created by using a Web Map Service (WMS). The data input and integration into the desktop GIS application can be facilitated by an instruction manual or a screencast (see Höhnle, Schubert & Uphues, 2012), which is worth considering in order not to put too much focus on the technical aspects of GIS. After the input/integration of the data, the spatial analyses are conducted using GIS functionality. During this analysis, the data, which are allocated to different layers, are examined regarding the hypotheses following a synthetically (re)constructive approach (Siegmund & Naumann, 2009, p. 4; see Fig. 10-2). This can take place by intersecting different layers or by performing combined queries (e. g. own measured noise levels and quality of building stock). This step leads to the composition of one or more maps, which can later be interpreted through hypothesis-testing. In this regard it also makes sense to test 'wrong' proposed solutions and consequently falsify them, which emphasizes the inquiring character of the approach. Subsequently, the results can be discussed critically and the hypotheses about noise exposure and its spatial effects can be preliminarily verified or falsified. Only a preliminary verification takes place because coincidences of this kind may suggest substantial relationships but do not actually prove them, nor even make assertions about the origin of relationships. However, the interpretation of the maps about noise exposure is not confined to the substantial level. Cartographic aspects, the (limited) validity/significance, and the potential for manipulation of the composed maps can also be critically reflected upon by comparing them to personal impressions of noise in the study area during data-collection.

(5) Noise and its possible spatial effects (correlations) can be visualized in the GIS; however, these are initially only assumed effects or relations. Spatial coincidences might be obvious, but they cannot be equated with causal relations. That is why these coincidences are used again for the generation of hypotheses. The hypotheses refer to the way the data seemingly relate to each other, that is to say that they address the actual causal relations between the factor of noise exposure and the dependent factors being examined. Furthermore, possible alternative explanations for the spatial structure of the dependent factors (e.g. rental prices) can be taken into account (e.g. property situation, spatial planning). This approach makes apparent the many factors that can play a role in the

manifestation of spatial structures. Thus, a more in-depth analysis is conducted, which goes beyond the visualization of spatial coincidences and shines light on underlying causal relations. During this step, a change of perspective on space may well take place. Up to this point, work has focused on an analysis following a largely spatial-analytical approach, which confines itself to physical-material space and at best has been enriched by individual impressions of spaces while collecting data. Now, the perspective is widened, the perception of space is included in the unit, and the focus can be put on residents' perceptions and actions in relation to the problem. This potentially deepens the use of the findings and encourages the critical questioning of them. At this stage, local residents may be surveyed about their perception of noise exposure, about how this affected their apartment-hunting, or if this might be a reason for them looking for a new apartment. In this way the individual as someone who constructs space becomes the focus of the teaching example. These results are again related to the previous results (the values from the noise measurements) as well as to the hypotheses. Thus, an inter-relation between objective, quantitative measured values and subjectively perceived, qualitative data on noise exposure emerges. This leads to a synthesis which illustrates the complexity of spatial processes. It becomes clear, for example, that spatial structures and processes cannot appropriately be explained simply by cause-and-effect relations (as assumed at the beginning of the module). The step is concluded by a discussion of the complete results on noise exposure and its effects.

(6) In terms of an inductive approach, the local results on noise pollution can subsequently be put into a larger context. In this case the focus should not be on general rules or models, but rather on counter-measures against noise pollution: personal results can initially be related to local noise control measures and they can also be observed for other cities in a further step. The European Environmental Noise Directive (2002/49/EC) offers an important frame of reference for this, which illustrates the EU-wide problem of noise exposure.

(7) As a last step, the whole project and inquiry process are reviewed and reflected upon. Ways of acquiring knowledge, the geographic perspective and its particular work procedures and methods can all be addressed. By means of this multi-perspectival approach, it becomes possible to question the validity and the limits of objective, measured data critically.

The teaching module presented here is spatially and thematically transferable. Its basic structure can be used for the analysis of other spatial phenomena. Information and material for teachers are presented below.

3 Preparation and Materials for Teachers

What is noise? Any form of sound which disturbs or harms people is called noise. Thus, it is very much subject to individual perception. Sound itself is a physical entity which is created by pressure variations in the air and which is diffused wave-like in the proximity. The resulting so-called sound-pressure level is measured in decibels (dB). How loud people perceive sound depends on the intensity and the frequency of the sound as well as individual personal factors. Sounds of mid-range frequency are usually perceived as louder than sounds of low range (deep sounds) and high-range frequency (high sounds). The resulting weighted noise level is usually measured in decibel A (dB(A)). Since this is a logarithmic measure, an increase of ten dB is actually a doubling of loudness. Street traffic is considered the main source of noise. The continuous flow of vehicles, as we can find on many of the main roads in the EU, can be seen as a source of noise visualized in the shape of a line. Sound propagation decreases for 3 dB(A) if the distance to the source of sound is doubled and if sound propagation is unimpeded (see Fig. 10-3).

Distance from middle of street in m	5	10	15	20	30	50	100	150	200
Noise level decrease in dB(A)	-7	-10	-12	-13	-15	-18	-22	-24	-26

Table 10-1: Decrease of noise level depending on distance from source of sound (middle of street)

dB(A) | Values of different sounds

Rustling of leaves	up to 40 dB(A)
Whispering	about 30 dB(A)
Breathing	about 25 dB(A)
Conversation	up to 60 dB(A)
Car (50km/h)	up to 80 dB(A)
Car (100 km/h)	up to 90 dB(A)
Jackhammer	up to 105 dB(A)
Jet airplane	up to 110 dB(A)

Materials. A device to measure the noise level in dB(A) or the sound pressure level is essential for the teaching module. There are professional devices available, but usually such devices are not available in schools as they are relatively expensive. For our module, smartphone apps, which are free of charge and measure sound using the integrated microphone, are perfectly suitable. 'Noise Meter' by the EEA is ideal. It is available for all common operating systems (iOS, Android, WinMobile). The app measures the environmental noise (in dB) for ten seconds, averages the value and categorizes it (good, mediocre, bad). Afterwards the result can made available to the EEA on a voluntary basis. The data are loaded into a database and visualized on a virtual globe (available on http://www.eyeon earth.org/). This crowd-based collection of environmental data can be viewed and analysed by scientists and private users. Another useful tool is the 'Noise Tube' app (http://noisetube.net/), which measures the noise level continuously and automatically geo-references the measurements. Once registered as users, people can download the data as a KML file, which can be integrated and edited, using Quantum-GIS for example.

Other materials required:

- Smartphone with GPS sensor and app for measurement of noise level (e.g. Noise Meter; alternative: sound level meter and GPS device)
- Large-scale map of study area (e.g. city map)
- Report sheets
- Hardware (PC etc.)
- Software (desktop GIS, Office, relevant apps)
- Further data (depending on question/hypothesis)
- Dictaphone (alternative: dictation app on smartphone)
- Questionnaires, writing utensils, solid writing pad

Finding an Appropriate Study Area. The area under investigation should be manageable, yet not too small. It should include areas with high as well as low noise exposure. These differences, or the spatial variation of the intensity of the factor noise, make it easier to analyse effects of noise exposure (analysis of correlations). It also makes sense to confine oneself to residential areas, because reference to the students' everyday life becomes more apparent in this way. Main routes, which pass through residential areas, are especially suitable, as often the main road is much used while the side roads are often traffic-calmed.

Measuring Sound. Reflecting on the preparation of the measurements is important. Students have to make up their minds about a variety of things: How great should the distances be between the points of measurement? Do we want to illustrate or measure noise exposure in points, lines (along the roads) or polygons/areas etc.? For the planned correlation analyses, a point measurement and illustration in lines are sufficient. It makes sense to divide up the routes into sections. For each of these sections, representative values are determined. If noise exposure is intended to be captured in polygons/areas, the measurement profiles have to be not only along possible sources of noise, but also crossways to them. Thus, a network of points of measurement emerges. Furthermore, the students have to make up their minds on how often to take a measurement at each point. Measuring noise at each point five times, with a time gap of fifteen seconds between and averaging the values afterwards, has proved to be a good approach here. When using the Noise Meter app, it is simply necessary to repeat measurements five times, with the same time interval between. In addition, a decision about the time of day for the measurements has to be taken (e.g. at night or at rush hour).

The following points must be considered in order to obtain reliable results (they apply to measurements using a sound level meter as well as to ones taken using a smartphone app):

- The device should be positioned in a way that the sound captured is received as directly as possible
- The most sensitive part of the microphone shouldbe directed at the source of the sound
- A wind shield for the microphone should be used in order to prevent sound-amplifying turbulence
- The microphone should be mounted at a height of at least 1.5 m
- A tripod should be used if possible
- The measuring device should be calibrated before and after each measurement
- As far as possible, data should be collected on calm days
- The GPS coordinates of the points of measurement should be saved or registered on a map
- The conditions at the time of measurement (date, time, weather) should be recorded.

Name(s): ________________				Date:					
Measurement	Coordinates	Time	Weather	Measured value in dB					
				.	.	.	.	.	
I									
II									
III									
⋮	⋮	⋮	⋮	⋮	⋮	⋮	⋮	⋮	⋮

Table 7-2 Sample measurement report

How can Noise be Illustrated using a GIS? The base data can be obtained from OpenStreetMap (OSM) using Web Map Services. Alternatively, OSM data can be integrated using the export function of the OSM website. The locally captured noise exposure can be depicted in a variety of ways: points, lines (noise level maps) or polygons (isoline maps). For school purposes, a visualizing noise-level map is usually sufficient. The line can be supplemented if necessary using the intersecting function. Thus, areas around the source where noise exposure is higher than a certain numerical value (see Table 10-1) can be indicated. In this way, it is possible to determine areas around the source of noise where the burden, computationally at least, exceeds a specific value (see Table 10-3). The rule of thumb that sound propagation decreases for 3 dB(A) if the distance from the source of sound is doubled and if sound propagation is unimpeded may be useful in this context (see Table 10-1). The following figures can help categorize noise levels:

	daytime	night-time
at hospitals, schools, spas and retirement homes	57 dB(A)	47 dB(A)
in general residential areas and small residential estates	59 dB(A)	49 dB(A)
in town centres, village areas and mixed-use areas	64 dB(A)	54 dB(A)
in industrial areas	69 dB(A)	59 dB(A)

Table 10-3: Emission limit values

In order to protect neighbouring areas from the harmful environmental impact of traffic noise, it has to be ensured that the emissions limits given below (Table 10-4) are not exceeded when building or carrying out essential changes or renovation.

Range	Evaluation
>70 dB(A) L_{DEN} *>60 dB(A)* L_{Night}	Very high exposure
65–70 dB(A) L_{DEN} *55–60 dB(A)* L_{Night}	High exposure
<65 db(A) L_{DEN} *<55 dB(A)* L_{Night}	Exposure/Disturbance

Table 10-4: Guidance for evaluation. (Source: Sixteenth legal ordinance to execute the German law for emission protection)

The L_{DEN} value is an average level for the whole year and describes the exposure in 24 hours (Day Evening Night). Calculating the noise value in the evening and night hours is weighted higher by adding five dB and 10 db respectively. The L_{DEN} value serves to evaluate general noise exposure. The L_{Night} value describes environmental noise at night for the whole year on average (exposure between 10 pm and 6 am) and serves to evaluate sleep disturbances. (Source: Lärmkontor GmbH, after Schiewe 2012 et al. [simplified])

The values measured can also be compared to official data. The EEA runs an online portal with examples of noise maps (http://noise.eionet.europa.eu/viewer.html).

The data of the dependent factors can be visualized by polygonal shapes (e. g. shapes of premises). Attributes can be added to them afterwards. In this context, a meaningful categorization of the values is important as it enables an easier analysis of relations.

Effects/Covariations to be Examined. The students should be able to identify numerous dependent factors. The following is a non-exhaustive list of factors which can be examined by the students:

- Rent level
- Land cost
- Vacancy (rate)
- Quality of building stock (possible categories: renovated/newly constructed, medium/mediocre, signs of decay)
- Inflow/Outflow of residents
- Type of building (possible categories: single-family home, duplex apartment, apartment block)

'Quality of building stock' can be an indicator for the financial status of the residents or the property owner. However, these indicators have to be critically discussed with regard to their validity and meaningfulness. During fieldwork, these dependent factors have to be further categorized if necessary. For example, students have to reflect on how a particular building can be assigned to one of the categories. For metric values (e. g. land cost, vacancy rate), students have to make up their minds about when a value can be considered high, medium or low. This makes it easier to interpret the data captured.

Further Exploration. Possible methods for closer examination are further surveys in the field. Students can develop questionnaires or conduct short interviews with local residents about noise perception or its effects on their everyday lives. The relations between noise exposure and its possible effects could then be analysed in this context. However, further causes for the spatial structures being studied (e. g. accumulation of single-family homes which are surrounded by apartment buildings) can be addressed here as well.

Possible Interdisciplinary Cooperation. The issue of noise affects many areas of human life, so it hardly surprising that it can be addressed in various subjects at school. For instance, biology can address the physical and deleterious effects of noise on the ear and brain more closely. Physics can put a focus on the generation and transmission of sound. Politics can address law-making procedures in the field of emission control.

Information material has been compiled using the following sources:

Amt für Umweltschutz des Kantons Zug (AfU): Lärmschutz. Straßenverkehrslärm. Zug. (= Online abrufbar unter: http://www.zug.ch/behoerden/baudirektion/amt-fuer-umweltschutz/laerm/strassenlaermsanierung-auf-kurs/laermschutz-strassenverkehrslaerm/at_download/file)

Fraedrich, W. (2005): 'Wie ermittelt man die Lärmbelastung?'. *Geographie heute* 26 (231/232), 45–47.

Nöthen, E. & Falk, G. C. (2004): 'LÄRM! Schüler erforschen mit GIS stadtökologische Phänomene'. *Praxis Geographie* 34, H. 2, S. 35–38.

Schiewe, J., B. Weninger & A.-L. Kornfeld (2012): 'Analyse und Verbesserung der Gebrauchstauglichkeit von Lärmkarten in der Öffentlichkeitsbeteiligung'. Dessau-Roßlau (http://www.uba.de/uba-info-medien/4304.html).

References

Anderson, R. D. (2007): 'Inquiry as an Organizing Theme for Science Curricula'. In: Abell, S. K. & Lederman, N. G. (eds.), *Handbook of Research on Science Education*. Mahwah, NJ: Erlbaum. 807–830.

Bette, J., Köhler, K., Miener, J.-P. & Miener, K. (2013): 'Lernen mit DesktopGIS im GU.' In T. Bartoschek & J. C. Schubert (eds.). Geoinformation im Geographieunterreicht. Grundlagen, Ptenziale, Unterrichtsideen. Münster: Universitätsverlag. 29-48.

European Parliament (2002): *Assessment and management of environmental noise – Declaration by the Commission in the Conciliation Committee on the Directive relating to the assessment and management of environmental noise*. Directive 2002/49/EC.

Europäische Kommission (2011): *Bericht der Kommission an das Europäische Parlament und den Rat über die Durchführung der Richtlinie über Umgebungslärm gemäß Artikel 11 der Richtlinie 2002/49/EG.*

Höhnle, S., Schubert, J. C. & Uphues, R. (2012): 'Die Entwicklung von Implementierungsstrategien für den GIS-Einsatz im Geographieunterricht - Ausgewählte Erkenntnisse eines Mixed-Methods Forschungsansatzes'. In K. Drieling, A. Hüttermann, P. Kirchner & S. Schuler (eds.), *Räumliche Orientierung: Karten und Geoinformationen im Unterricht*. Braunschweig: Klett. 308–318

Siegmund, A. & Naumann, S. (2009): 'GIS in der Schule. Potenziale für den Geographieunterricht von heute'. *Praxis Geographie* 39, H. 2, 4–8.

CHAPTER ELEVEN

MAKING MY PLACE IN THE WORLD: USING GEOMEDIA TO EXPLORE PLACE AND COMMUNITY

SUE BERMINGHAM AND JOHN LYON

Abstract

The following set of activities were developed within the context of the English geography curriculum at KS4 (14- to 16-year-old students) for a project funded by the Paul Hamlyn foundation. The project, entitled 'Making My Place in the World', worked with four secondary schools, exploring ways to encourage young people to play a part in helping shape their community as active citizens by participating in real-world conversations about the places in which they live, and through these conversations both develop their speaking and listening skills, and increase their confidence, self-esteem and awareness of themselves as participants and agents of influence in the world.

These young peoples' spatial lives are shaped by powerful local and global forces. We aimed to develop their understanding of these forces by drawing on the knowledge, interests, narratives and concerns that emerge from their local, digitized everyday life, and from that beginning, to work to construct new knowledge and understandings. Integral to this was the use of a range of geomedia. Geomedia were used to develop deeper understanding of the concepts of place and space, how we view place, and how different people may think about the same place in entirely different ways.

1 Aims, Target Group and Pedagogical Background

Aims:

- To encourage young people to play a part in helping shape their community as active citizens.
- To contribute to the Digital Agenda for Europe 2020 – 'It is essential to educate European citizens to use ICT and digital media'

Target Standard and Target Group. UK GCSE (KS4 aged 14–16) – Class set of 25 students studying geography.

Pedagogical Background. The overarching aim of the work was to develop a conceptual framework for interrogating places and spaces. We intended to explore the concepts of place, space, scale and change through the context of the daily lives of the pupils. This involved the need to develop a richer vocabulary for talking about familiar places and 'everyday geographies'. It was apparent that promoting conversations between pupils and place managers required all parties involved to examine and develop the use of complex vocabulary. Planners and place managers have a level of discourse that may act as a barrier to pupils if it is not fully understood. Dialogue with planners at a professional level contributes to increased confidence with speaking and listening skills. Teachers and pupils were able to develop cross-curricular links through literacy and English. The learning was underpinned by an enquiry approach, with the pupils using geomedia to analyse and solve spatial problems.

2 Description of the Learning Environment

The project funded the appointment of two Community Geographers (CGs), for one day per week, to support the teachers in their work. The CGs' role was wide ranging, involving activities such as negotiating and setting up links with planners and other adults with responsibility in the local area, in addition to devising activities, and supporting and teaching in class. A varied sequence of lessons was planned for Y9 and Y10 classes (ages 13–15) in four schools in two cities in the UK, namely Sheffield and Greater Manchester. A wide range of activities were carried out, such as brokering meetings with adults, for example planners, developing and disseminating ideas, in addition to a range of lesson activities.

Geography is the ‘world subject’: it helps us develop knowledge and understanding of the world, and our place in it. Through its study we develop the knowledge, understanding and skills that enable us to grasp our common humanity and our relationships with each other and the environmental resources on which life depends. By thinking geographically about the past and the present, we are better equipped to imagine our possible futures. However, skills of any sort are worthless if one has neither the inclination nor the good sense to use them. Key aims were to help young people develop the general disposition to think ‘better’ within the context of their own place as they perceive it, to locate places spatially, and to understand the key processes of change. Geography was chosen as the subject area to set up the teaching as geography as a discipline gives us the conceptual sophistication to interrogate places and spaces and has close links to politics, citizenship and wider global issues.

The project activities explored how young people view their world, how they talked about their places, and how their actions may have impacts on their places. Students were encouraged to participate in real-world conversations about the places in which they live, and through these conversations develop their speaking and listening skills, and increase their confidence, self-esteem and awareness of themselves as participants and agents of influence in the world. The activities were also designed to help them to understand that places may be seen differently and have different meanings both at different times and to different social groupings. We considered what sense of belonging young people have to place(s) and what ‘local’ means to them.

The learning encouraged young people to work with each other, their CG, and with local decision makers, such as planners, residents and others from their local community, on a particular issue the students chose themselves. The intention was to encourage young people to identify and investigate their own issues of concern, perhaps focusing on an area of significant redevelopment that may have polarized opinion in the community.

Initial Activity: A Values Continuum Exercise. The CGs wanted to find out a little about the students, establish a relationship with them and explore their perceptions of place – places that might be special to them, that they might want to learn more about. The project also wanted to develop the idea of the importance of connecting young people to the people who control and manage places.

At the outset, an aspiration of the project was that the young people would wish to focus on an area of redevelopment or an inequality issue that had polarized opinion in the community, but it is important to remember that what adults see as relevant may be viewed from a different perspective when our assumptions are replaced with evidence of how young people experience and construct their worlds. The project team wanted to allow student voices to be dominant in identifying the key issues. The project hoped to develop awareness that theirs, the students', is only one view among a range of views, and an understanding of and respect for diversity. The project also hoped that, through effectively developing their speaking and listening skills, students would reach a better understanding that they can have a voice that contributes to the future of their local area. At the end of the project, one student commented 'What I've learnt is that you don't really know a place until you've talked to other people to see what they think about it.'

The CGs and teacher asked a series of questions, and in response to each question students were invited to place themselves along the scale (see Fig. 11-1). The team used the classroom to frame the scale, i.e. from one end of the classroom to the other. Students placed themselves in a line across the classroom according to their agreement/disagreement with the statements.

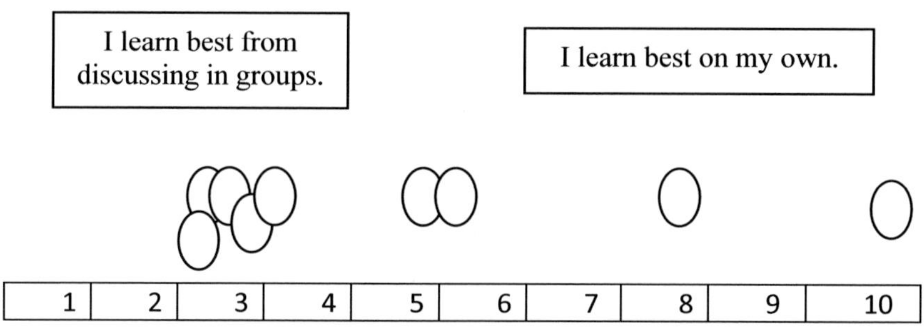

Figure 11-1: Opinion Line (Scale of agreement/disagreement, where 1 = completely agree and 10 = completely disagree)

The teacher invited students to give the reasons for their position along the line.

Other sample statements for students to respond to using the scale:

- Teachers do not see what pupils see
- The best lessons are active

- When I leave school I want to go to University
- I intend to live in this area when I am an adult
- I am proud of my local area.

The students and teachers were also asked to contribute their own statements.

Values Continuum Exercise using Images of Place. Each student was given a different image of the local area and asked: *Does your image represent your place?* The language of what the project means by 'represent' needed to be discussed with pupils.

Students were again invited to place themselves along the opinion line (from one end of the classroom to the other) according to their agreement/disagreement with the statement. (See Fig. 11-2.)

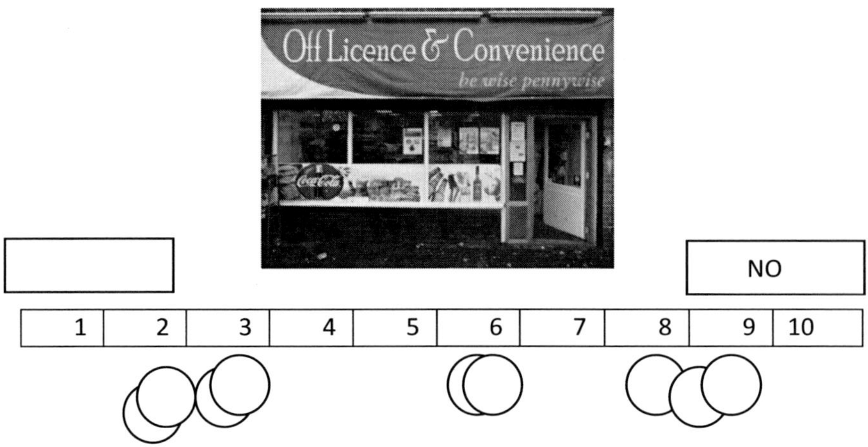

Figure 11-2: Opinion line (Scale of agreement/ disagreement).

Students were then asked to describe their image to the others and why they felt it did or did not represent their area, such as 'What can't we see or know from the photo?'; 'Can you see "change"?'; 'What do you think other people think of your place?'. As they discussed the views and images, the CGs found that the students had very limited vocabulary to describe photographs.

Developing a Language for Talking about Place. The project team decided to agree upon and use a language for talking about place. This was not about speaking 'properly' or 'correctly', but sensitively and appropriately. The project wanted to develop a kind of speaking that is *adaptive*, *varied* and *sensitive to context*, and a kind of listening that is characterized by *evaluative participation* and *sensitivity* to the development of discussions.

The project identified words that planners might use during their normal workday. It also looked at words needed to engage with the urban topics on their GSCE specifications and developed a vocabulary list.

The words were taken from the Sheffield Planning Website and turned into a Word Cloud using *http://www.wordle.net/* (see Fig. 11-3).

Figure 11-3: Word Cloud for Sheffield City

The project team organized a series of activities to introduce the students to appropriate language skills and the vocabulary needed to engage with planners and ultimately with the GCSE exam. It soon became apparent to the students that they needed to build an appropriate vocabulary as well as hone their skills of observation, speaking and listening, so that they could adopt a more formal style of talking, depending on their audience. Using a previously compiled wordlist, both the CGs ran a range of activities using literature, images and discussions to embed the words into the students' vocabulary. Some words were distinctly geographical, such as 'gentrification', while others such as 'congested', rather than 'busy', extended their general vocabulary.

Identifying Issues. The final choices for study in the four schools were as follows: to explore place and change along an arterial road into the centre of the city; to understand the location and purpose of CCTV cameras at a local shopping centre; the regeneration of the town centre, and the refurbishment of an iconic block of flats in the city.

Virtual Fieldwork – Using Google Maps and StreetView. The project team set up a blog using WordPress (a free and easy-to-use service) for 2 separate areas of Sheffield. Project pages were created in Google Maps for two separate areas.

The project page in Google Maps (see Fig. 11-4) enabled the CGs to do the following;

- upload and geo-reference photographs
- upload and geo-reference the planners' podcasts
- upload and reference DVD clips
- add files showing student tracks

Figure 11-4: Project page in Google Maps.

See examples of each page at:
Blog: http://shareit.yhgfl.net/yhgfl/ga/
Project 1: http://bit.ly/KwUksK
Project: 2 http://g.co/maps/kj2x3

The project team chose to create virtual fieldwork environments using Google Maps to enable students to access the area virtually prior to a real visit. One major value of using Google Street View was that it enabled the students to get a real feel for the area and 'walk' through it, making use of the 360° images on StreetView. More recent images which were geo-referenced and located on the map were also added. In order to be able to do this, all the materials were first uploaded to an online site from where they could be retrieved and then added to Google Maps.

Students were able to open the very recent geo-referenced photographs and compare them against historical images of street views to explore change, decay and regeneration. The blog allowed students to discuss issues and their ideas about the benefits arising from regeneration and change.

The project team realized that in order for the students to increase their own vocabulary, it would be beneficial for them to be able to listen to planners talking about planning issues and using specific planning terms and vocabulary. A series of podcasts was recorded by the City Planners using very simple technology. These were uploaded, geo-referenced and added to the project page. Students were then able to listen to the planners describing and explaining changes while looking at either the 360° images or the more recent images.

The planners also provided historical images in a digital format and excellent, professionally-made video extracts, which were also uploaded to the site.

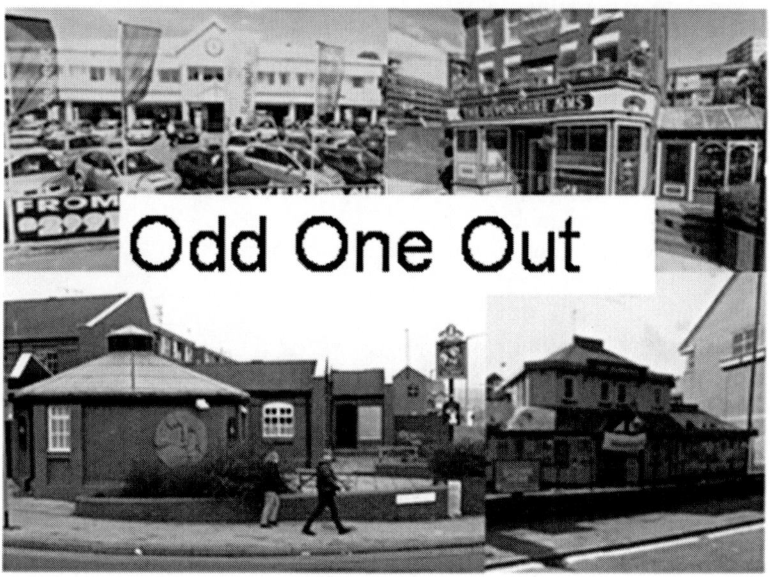

Figure 11-5: Google Street View images

There are a number of ways in which this virtual digital landscape and the images can be used. The students were able to take a virtual walk through the landscape exploring change through listening to the planners' podcasts and comparing Google Street View images with the newer geo-referenced images taken by the CGs. In one exercise, students were asked to examine four images taken from Google Street View. The task was to identify the odd one out (see Fig. 11-5).

This led to some interesting discussion, prompting a range of answers. The teacher looked for students to identify the pub in the bottom right-hand corner of the image as the correct answer. All the other images show buildings that have been demolished since the Google Camera car passed. The students were able to confirm the differences between the old and new cityscape by opening the more recent geo-referenced photographs, as discussed earlier. This exercise demonstrated how much change had occurred in an area where the students had not initially noticed any. The city planners are aiming to change the image and coherence of this area within the city and were keen to understand the students' impressions of the changes and their thoughts for the future.

Following initial virtual investigations, the students were taken to the study areas to gain a greater awareness and understanding.

Some students were issued with tracking devices (some used a mobile phone App) to record their journeys and the route of their field visit. Students were asked to think about planning change and to talk to local people about benefits, impacts and concerns. Using the trackers they were able to geo-reference any images that they felt represented their findings. They were asked to take images that might be:

- A view for a geographer
- A 'Wow' photograph
- A photograph which captures human involvement
- A picture that shows nature
- A picture that tells a story
- A picture that a child would take
- A picture that might make someone cry
- A picture which best represents this place for you
- A picture of something unusual, funny or whatever – you decide.

The example image (see Fig. 11-6) (created using the Hockneyizer at Big Huge Labs) could be one way of showing how a place might be pictured.

Figure 11-6: Example image of how a place might be pictured

Using iPads issued to students and a mobile WiFi, the students were able to open the project pages and listen to the planner talking about proposed changes while looking at the cityscape being described.

Presenting the Findings. Students were asked to produce a PowerPoint presentation of no more than 4 slides presenting recent and future planning proposals, including benefits, impacts and concerns. The audience they worked towards was a local community group. Students were asked to use photos, screen dumps, Internet sites and the planning podcasts to help them.

In order to produce the PowerPoint the students needed to complete the following activities:

- Click on the markers denoting the planners' podcasts about planning and change
- Click on the markers denoting recent photographs and compare with images on Google Street View to identify recent changes
- Save any appropriate photos to use in their presentation
- Check the Internet for extra information.

3 Preparation for the Teacher

Freedom. The first decision for the teacher is to decide how much freedom to give to the students in choosing their study area. This largely depends on the teacher and her/his desired outcomes from the project. Students can be given entire freedom of choice or the teacher may consider delimiting the study area and choice of topic.

Resource Preparation. Some images and sound files etc. need to be made before the project begins. Others can be added later by the teacher or students, depending on the project's purposes. All images, sound files etc. need to be uploaded to the web for storage and access. For this project, we used a government-sponsored site to host files, but there are a number of free services on the Internet.

ICT Experience. Experience in using and creating project pages in Google Maps is essential. All images, sound files etc. need adding and geo-referencing on Google Maps.

Equipment. The fieldwork equipment used can vary, depending on the project funding. We bought two iPads, which students were able to use in class and in the field. We also purchased a simple recording device and headphones with speakers to enable reasonably high quality sound recordings. The project funded schools to purchase cameras and tracking devices for students. In school, the teacher needs access to PCs or laptops that are Internet-linked to enable the students to work individually or in pairs. Other devices can be bought if the funding allows, such as simple GIS trackers, video cameras and sound recorders.

4 Experiences

Overall the experience was very positive. The students were motivated and challenged, but interestingly quite a number did not know how to use Google Street View with any confidence. While we did ask students if they felt confident with Google Street View and all claimed they were, in fact a number did not know how to access the 360° image in Street View. This caused some issues in the first lesson as a number of students needed help. The sound levels in the town planners' first recordings were low, and although they could be heard from a PC when tested at home, it proved difficult to hear them in the noisier environment of the classroom. Some recordings needed to be re-recorded to ensure clean, clear sound. We ran a similar project in two schools with students aged between 13 and 14. The project proved successful in both schools and with mixed ability groups in each. Both schools intend to take the work from the project to embed a unit into their curriculum for the next academic year.

References

Bright, N. G. (2011): "'Off the Model': Resistant Spaces, School Disaffection and "Aspiration" in a Former Coal-Mining Community'. *Children's Geographies* 9 (1), 63–78.

Brooks, R. (1996): 'Creating a Positive School Climate: Strategies for Fostering Self-Esteem, Motivation, and Resilience'. In J. Cohen (ed.), *Educating Minds and Hearts: Social Emotional Learning and the Passage into Adolescence*). New York: Columbia Teachers College Press. 61–73.

Castree, N. (2009): 'Place: Connections and Boundaries in an Interdependent World'. in N. J. Clifford, S .L. Holloway, S. P. Rice & G. Valentine (eds.), *Key Concepts in Geography.* London: Sage.

Emmel, N. & Clark, A. (2009): *The Methods Used in Connected Lives: Investigating Networks, Neighbourhoods and Communities*. ESRC National Centre for Research Methods, NCRM Working Paper Series 06/09.

Emmel, N. & Clark, A. (2010): *Using Walking Interviews*. ESRC National Centre for Research Methods, Realities Toolkit 13.

Katz C. (2009): 'Social Systems: Thinking about Society, Identity, Power and Resistance'. In N. J. Clifford, S. L. Holloway, S. P. Rice & G. Valentine (eds.), *Key Concepts in Geography*. London: Sage.

Thrift, N. (2009): 'Space, the Fundamental Stuff of Geography'. In N. J. Clifford, S. L. Holloway, S. P. Rice & G. Valentine (eds.), *Key Concepts in Geography*. London: Sage.

Todd R. H. (2007): 'Place-Based Learning in Teacher Education: A Windshield Survey'. *Social Studies Research and Practice, Texas Tech University* 2 (3), 360–402.

CHAPTER TWELVE

USING MOBILE DEVICES AND GEOMEDIA FOR ANALYSING THE IMPACT OF NOISE IN AN EVERYDAY STUDENT SPACE: THE PLAYGROUND

JÉRÔME STAUB

Abstract

This experiment is part of the geography and civics curriculums of a secondary school in France. It aims to analyse the soundscapes of selected student spaces, especially the playground, and to be aware of the noise levels and noise nuisances of this space. It relies on the use of mobile phones as sensors to produce different sound maps, mixing soundscapes, ways of listening, and the measurement and perception of sound levels. These mobile sensors were used to query Sharples's theory (Sharples, 2007) on the use of mobile devices in education.

1 Introduction

According to a survey published in 2010, two out of three French people say they are bothered by noise at home. The impact of noise on French lifestyle is considerable. In this context, education should perhaps have a mixed mission: to encourage students to be aware of noise pollution, and to develop their learning by listening to noises and sounds. The main argument is to show how the perceived noise level of a space depends on the way people are listening to and experiencing the sound environment. Several surveys have identified three main areas of noise in schools: the cafeteria, the classroom and the playground. The third of these is the subject of this experiment. It offers a particular pathway into the analysis of space, based on the study of soundscapes. In this project, this feature is coupled with the use of mobile phones. In fact, the experiment interrogates the school space in different ways: it wishes to highlight the relevance of soundscapes as a source for studying space; it looks at the effectiveness of the use of mobile phones and their consequences; and it examines the difficulty of reconciling measured data and sensory data.

Pedagogical Background, Aims and Target Groups. The choice of the appropriate survey site was almost self-imposed. This was in part because of the findings of the students, but also those of the teachers (also suggested were the school cafeteria and the classroom). The playground is the place where students create soundscapes in a variety of different ways, such as their own voices and the music or videos they play on their mobile devices. Interestingly, the students never acknowledge hearing the original background sound environment, though the impression of musical cacophony is ever present when crossing the playground. The playground is a place where students spend considerable time each day, but without paying attention to how the space is structured.

On average, students spend one hour a day in the playground, approximately 180 hours in the year. This is not a trivial or anecdotal place. Moreover, time spent here is not only statistically significant but also relates closely to cultural and sporting dynamics. This is an area of the non-formal school environment, part of the everyday spaces of students.

These findings should be placed in parallel with the explosion of students' personal digital equipment, such as mobile phones. According to a recent survey (http://www.netpublic.fr/2012/03/pratiquesnumeriques-jeunes-2012-presentation/), 80% of 11- to 14-year-olds have a basic or mid-range mobile phone. In most cases, this daily tool is rarely or never used in an educational context. In general, the use of such devices is prohibited in the school building and grounds.

The students selected for this project are from two grades: *sixième* (first year of secondary school; students aged 10–11), and *cinquième* (second year of secondary school; students aged 11–12). The choice of these levels was in part determined by the desire to develop the study over several years.

It was also a strategic choice, as this work is easier to realize for the first time if the students have less experience of the secondary school structure.

Moreover, the choice of these grade levels also corresponds to the curriculum guidelines set by the National Ministry of Education. Thus, for the younger ones, this process of discovery of the school establishment refers to two themes, in geography and civic education. In geography, the curriculum refers to the discovery and analysis of the school neighbourhood.

Soundscapes can be used to initiate the study. In civic education, the first element of the curriculum is centred on the school and the student. For older students, sustainable development is the central theme of the geography programme; noise pollution can be addressed in this context.

Background. This work is at the crossroads of several studies involving four elements: the soundscape, the living space, the significance of noise in school spaces, and using mobile devices in an educational context (Figure 12-1). The theme of the soundscape refers to the pioneering studies of Raymond Schafer, which show how it is possible to define a soundscape:

> I call the acoustic environment the soundscape, by which I mean the total field of sounds wherever we are. (Schafer, 1992)

He was one of the first to propose a soundscape map. The term was challenged by Paul Rodaway:

> Though Schafer himself is a musician and composer, his terminology is heavily reliant on visual metaphors. (Rodaway, 1994:84)

Rodaway himself replaced the term 'soundscape' by 'auditory geographies', which make greater use of the the concept of active listening. In this approach, the soundscape concept is related to the listening environment in a much more reflective manner. Moreover, spaces are studied in terms of the analysis of relative spaces and lived spaces. Guy Di Meo defines the lived space:

> The living space for every individual merges with the area of spatial practices. It corresponds to the frequented area. […] The living space reflects a practical experience of the place. Every practice has a dimension of human imagination. (Di Meo, 1988)

According to Sharples, Taylor and Vavoula (2007), mobile devices have significant potential for educational activities in field surveys, integrating a scientific approach. In order to understand and analyse mobile devices for learning, Sharples et al. created a mobile learning model (Fig. 12-2).

When mapping soundscapes, as well as when using the concepts of lived spaces, the present study questioned Sharples's model concerning the various uses of mobile tools, in order to show the relevance and limitations of such tools in educational situations.

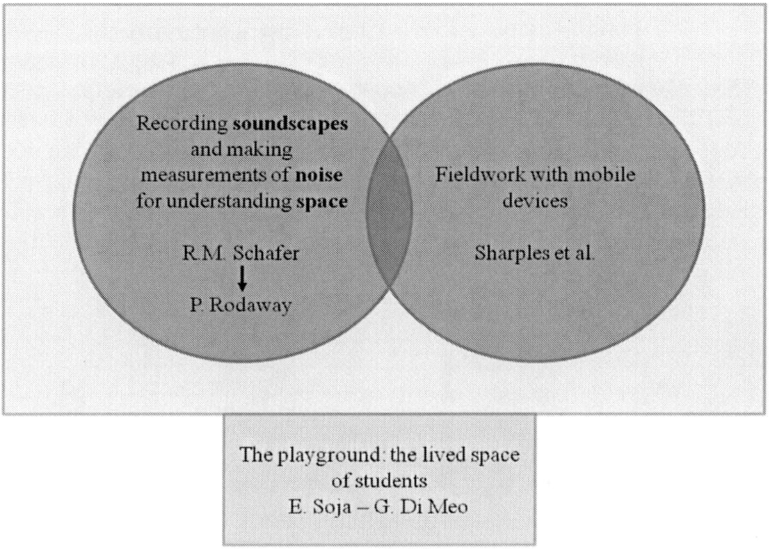

Figure 12-1: Relations between experiment and background

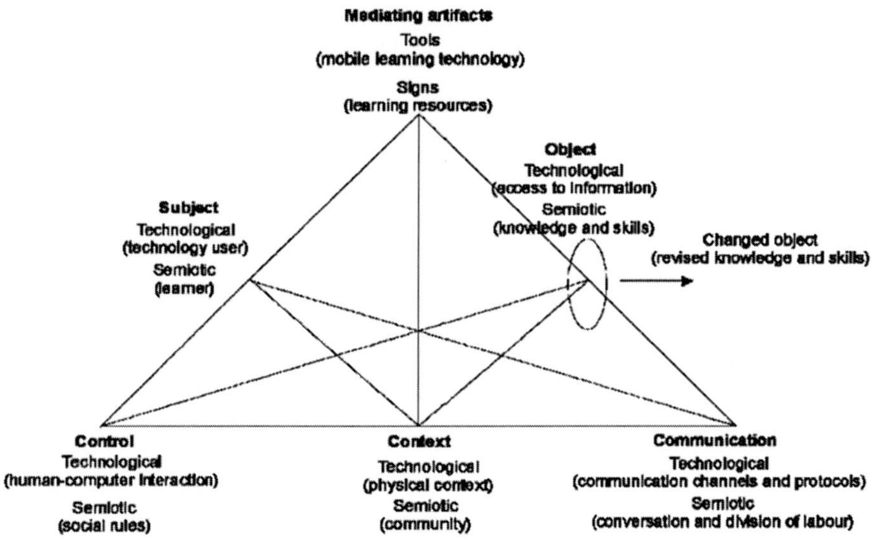

Figure 12-2: A framework for analysing mobile learning (Sharples, Taylor & Vavoula, 2007)

2 Description of the Learning Environment

The playground remains a complex space to analyse, for several reasons. First of all, this is the space most often frequented by students, and the one concerning which we have most stories and emotional experiences. In addition, sound variations are very significant, depending on the ways the playground is being used. Therefore, this space requires special and progressive analysis. Thus, we organized this experiment into several steps (Fig. 12-3).

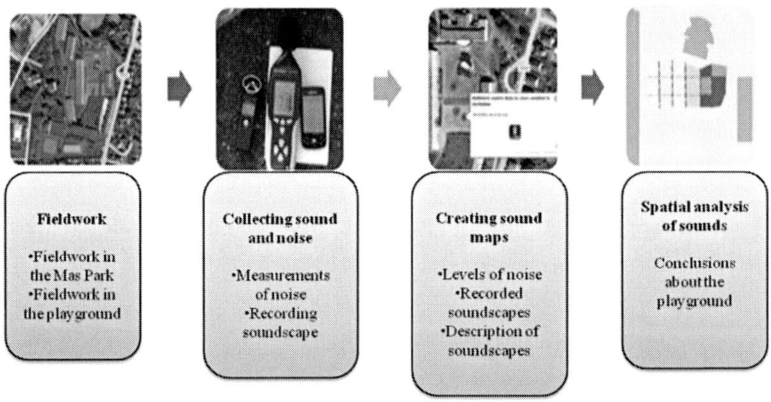

Figure 12-3: Description of the whole process

Step One: Use a Familiar Space for the First Exploration of Soundscapes. As a space, the playground will be very familiar to students and provides a suitable closed location for a first trial approach to soundscape evaluation and for practising listening.

For an hour, the students practised actively listening to the soundscape and made measurements of the noise levels. Initially, the teachers used the following activity: for one minute, everybody listened to the sounds around them and noted on paper the origin of each sound heard. Each student then expressed what he/she had heard. The teacher then provided a summary of all the sounds collected to form the sound environment.

A second activity was designed that made use of the following equipment: a digital recorder, a sound level meter, and smartphone apps (e.g. Noisetube: http://noisetube.net/?locale=fr; WideNoise: http://www.widetag.com/widenoise/). Each group of 4 or 6 students made measurements and recorded sounds over a period of five minutes in a place of their choice.

The third activity was to work in groups in different places in the school vicinity. Each student had a clearly defined task within the group. One person worked on listening, indicating on a plan the different sounds heard, and their directions using arrows of varying thickness, depending on the perceived intensity of the sound. A second person recorded the sound 'atmosphere' (with a mobile phone and/or a digital recorder); a third student used a sound meter to measure the noise levels. The exact time of survey was logged by each member of the group, and the activity was carried out in two or three other places in the school vicinity.

Step Two: Collecting Sound Data in the School Playground. All the experiments were carried out during one afternoon, with one being undertaken during the recreation break. Developing a more scientific distribution of surveys in the playground space was easier because it was a closed space. The identification of landmarks like benches and trees makes it easier to compare the measured intensity of the sound and the perceived intensity. For the latter, we asked students to characterize the perceived sound intensity of a space using a score out of ten (Figs 12-7 and 12-8).

Step Three: Cartographic Exploitation of Fieldwork Results. The mapping of data collected in the field was the most important part of the project. ESRI solutions were used: ArcGIS Online and ArcGIS Explorer Online. Three different approaches to mapping the data sets gathered were presented: mapping orientated to the ambient sounds heard and recorded; mapping of the sound intensity using the NoiseTube app; mapping of the perceived intensity.

The first approach involves multimedia mapping: the recorded sounds are integrated as a playlist in an online sound database which works like a social network – AudioBoo (http://audioboo.fm/). The information created included a description of the atmosphere as well as a link to the online recorded background (Fig. 12-4).

The second mapping activity was to translate the data sheets of listening information, including the orientation and the origin of the sound (Fig. 12-5).

Figure 12-4: The sound map of the playground

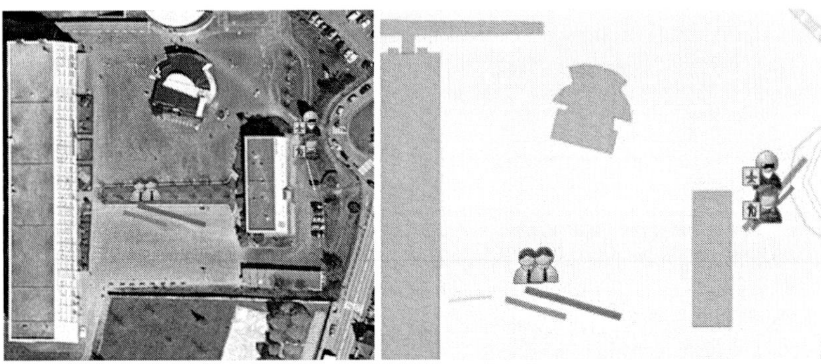

Figure 12-5: The heard sound map (Bing Maps and Open Street Map)

Finally, work on sound intensity was based on two types of mapping: one corresponding to the integration of the data collected, and geo-tagged by the NoiseTube app; the second mapping activity, more sensitive and difficult to implement, was to transcribe the perceived intensity scores out of 10. The solid colours corresponding to the scale of the sound intensity of the NoiseTube app were close to those proposed by the Public agency for the environment and energy (ADEME9) (Figs 12-6, 12-7, 12-8, 12-9).

Figure 12-6: The NoiseTube Map

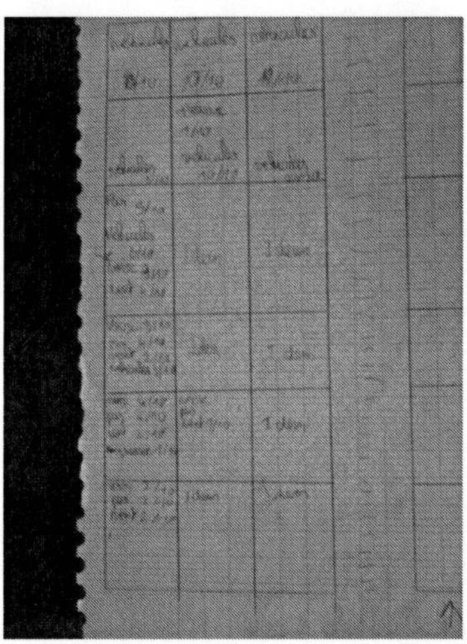

Figure 12-7: The perceived noise data collected by the students
Figure 12-8: The perceived noise map (In Bing Maps)

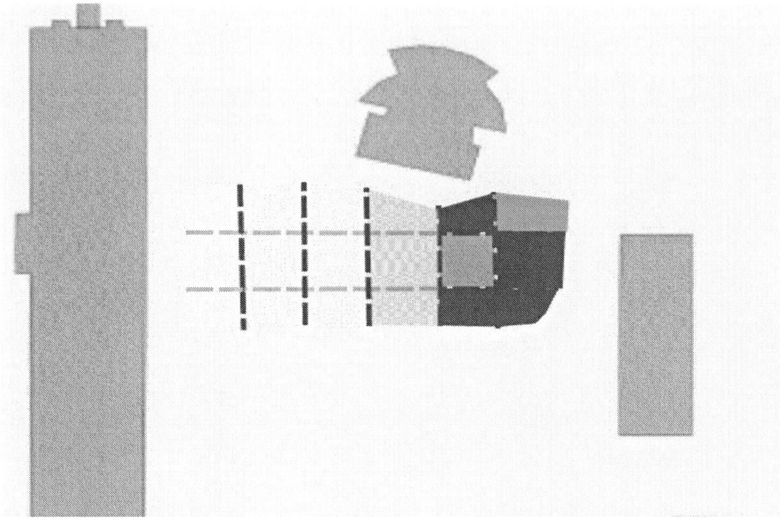

Figure 12-9: The perceived noise map (in OpenStreetMap)

Step Four: Spatial Analysis of Sounds. The analysis of the sound data in relation to space gave rise to several conclusions.

Except during school breaks, the playground has a sound atmosphere comparable to that of a typical urban park: the natural sounds (birds, leaves in the wind) are constantly present, accompanied by background noise such as road traffic. This was a true discovery for the students, who previously saw the playground as a noisy space for communications. But when the playground fulfilled its role of recreation, listening and taking readings of specific sounds weren´t possible because of the large number of loud voices heard. Only one witness record was carried out during these times. Because of this inability to collect information, the students gained a new awareness of the sound intensity of the playground at these times and also started to become aware of different sounds. The students discovered that they themselves were largely the origin of the noise of this space.

As for the readings of sound intensity, the comparison between the actual measurements and the perceived intensity shows some differences.

The perceived results exaggerate the actual sound levels. This is consistent with different surveys carried out in many urban agglomerations in the framework of the "Plan for the Prevention of the Noise in the Environment" (PPBE) producing noise maps that conform to European regulations.

3 Evaluation and Transferability

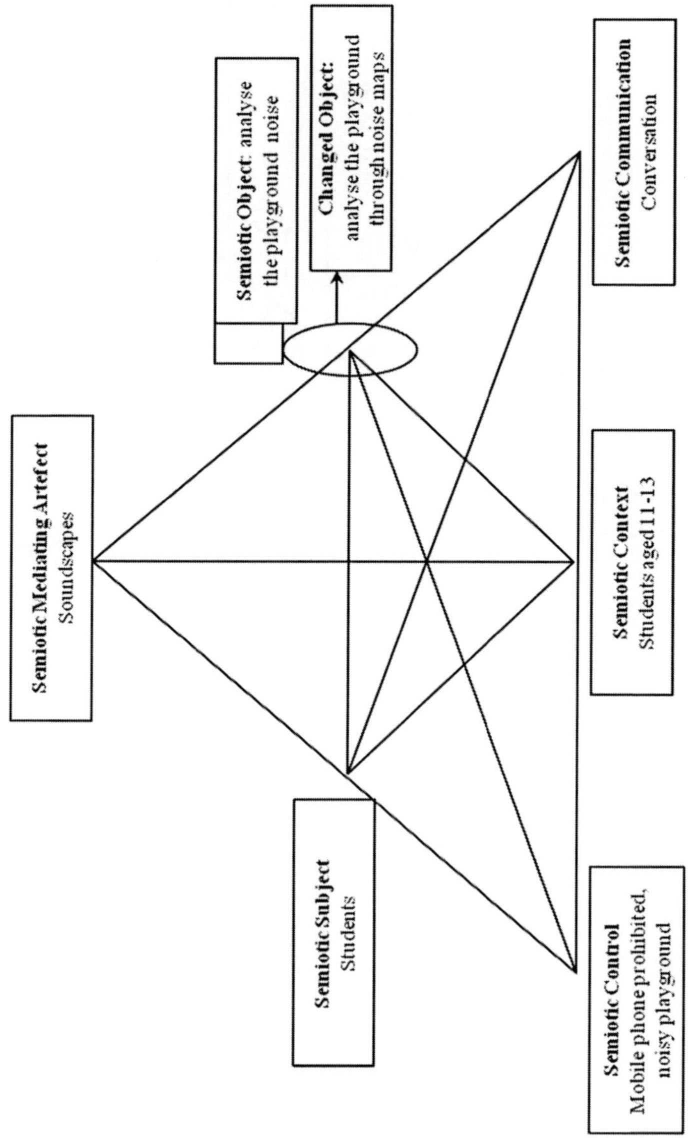

Figure 12-10: The Semiotic View (Sharples, Taylor & Vavoula: 2007, 238)

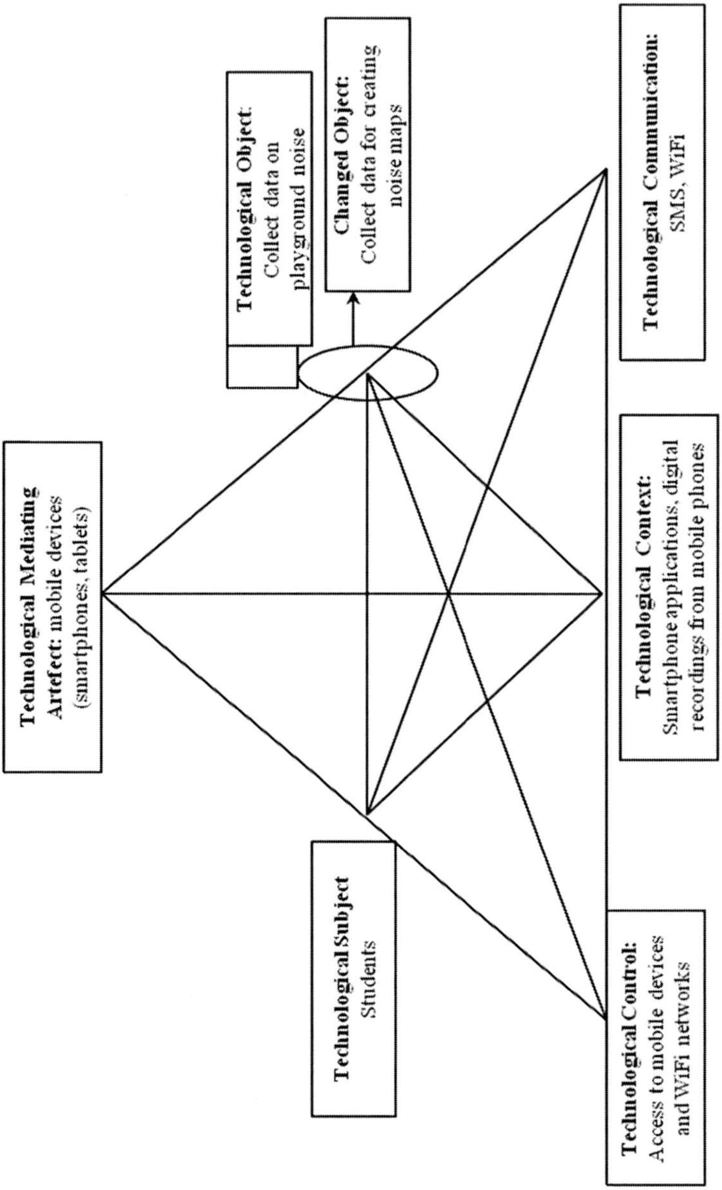

Figure 12-11: The Technological View (Sharples, Taylor & Vavoula: 2007, 239)

Mobile Devices in Education. If we examine the impacts of this experiment from the standpoint of the students in this case study, set within the the context of elements from Sharples's learning theory in the mobile age, we see that the outcomes connect two levels of cognition, the semiotic view and the technological view:

A comparative analysis of these diagrams illustrates the important changes brought about by the use of mobile devices, in particular smartphones. These changes include the new possibilities and availability of digital mapping functions.

The use of GPS involves the transfer of data to an online platform via WiFi networks. The other indicators, identified and mapped in a subjective form, complement this work in the context of web mapping. In fact, the most complex work was to transcribe the subjective data collected in situ (listening patterns, recordings) into digital form, and then to upload the data into the mapping matrix.

In parallel, the use of mobile tools requires a permanent wireless connection, which was not the case for all of the locations visited. Nor do all applications allow data to be retained in a cache, which makes them difficult to use, despite their apparent attractiveness.

Consequences for the Student's Spatial Skills. The diagrams indicate the importance of the use of digital mapping as a means of supporting analysis. These digital activities greatly alter the spatial skills of students.

They become the central actors in the experiment and especially in data collection, cartographic production, and spatial analysis. This centrality of the student in the whole learning process gives students greater autonomy and empowers them as citizens to reconstruct such an approach, applying it to other spaces or topics. In addition, the connection between the different phases of implementation gives students an opportunity for critical evaluation, both of the scientific approach adopted and of the results obtained. The students see maps not as a full scientific reference, but as subjective cartographic constructions. This subjectivity is also expressed in the delicate operation of translating sensitive data into cartographic and digital (and therefore to some extent mathematical) languages. This work remains imperfect and corresponds to the human reaction of modelling and coding of the surrounding world.

Transferability. This work offers several possibilities of transferability. The theme of sound in schools is part of a European movement of sound analysis (including through an EU Directive). It can also be a separate entry in the educational paradigm of sustainable development. Spatial skills that are mobilized by this experiment are directly related to the student in the 21st century, as defined by UNESCO. It is perfectly possible to imagine a similar project focusing on a very different space, using the same methodological framework. In fact, the work presented here offers several types of portability, skills acquired and spaces studied.

4 Conclusion

This research with sound maps shows how listening to the environment can become an important element in understanding a space, in particular the familiar space we live in. These areas are marked mostly by cultural, mainly visual, perspectives. This study is also an interesting educational lever to change student perceptions of space, even though this might be ephemeral.

The involvement of mobile tools also modifies the support needed: because the data are geo-tagged (whatever their nature: text data, images, sounds, videos ...), they require the use of digital mapping. In fact, the mobile device becomes a scientific toolkit for students, although the data collected may lack details. The transfer of data from a mobile to a web interface facilitates the critical analysis of information. The transcription of more sensitive data remains more complex, but also more interesting as the student realizes the potential of creating maps.

In this specific case, an indicator combining sound level measurement and the intensity felt could have been created, using a simple mathematical model, possibly offering a gradation of solid colours.

References

Di Méo, G. (1998): *Géographie sociale et territoire*. Paris: Nathan.

Rodaway, P. (1994): *Sensuous Geographies: Body, Sense and Place*. New York and London: Routlege.

Schafer, R.M. (1992): *A Sound Education: 100 Exercises in Listening and Sound-Making*. Indian River, Ontario: Arcana.

—. (2010): *Le Paysage sonore, le monde comme musique*. Paris: Wildproject.

Sharples, M., Taylor, J. & Vavoula, G. (2007): 'A Theory of Learning for the Mobile Age.' In R. Andrews & C. Haythornthwaite (eds.), *The Sage Handbook of E-learning Research*). London: Sage. 221–247.

CHAPTER THIRTEEN

'SEE YOU': A WEB-BASED APPROACH FOR TEACHING ABOUT GPS, MAP ANALYSIS AND PRIVACY TO SECONDARY SCHOOL STUDENTS

HANS-JÖRG STARK

Abstract

GIS in teaching and the promotion of spatial technologies in secondary schools have been very lively in recent years. The 'See You' project is a successful representative of this movement. It provides an easy-to-use web-based application that allows for both students and teachers to monitor, investigate and discuss their spatial leisure behaviour based on their recorded GPS tracks. It also fosters an awareness of the opportunities and risks associated with the modern tracking technologies that are found in smartphones and similar devices, and sensitizes students to the issue of privacy.

1 Pedagogical Background, Aims and Target Groups

The School of Teacher Education at the University of Applied Sciences and Arts, Northwestern Switzerland (FHNW) wanted to support the interest of secondary school students in technical disciplines. Hence they started an initiative with the goal of creating special lessons and tutorials relating to four selected domains. These lessons were prepared by an interdisciplinary research team and were provided to teachers at secondary school level for free. One of the four areas was the field of Geomatics, for which a pre-excisting secondary school project, 'Map Your World', was chosen. The initiative allowed a time budget of only four or so lessons for each of the four technical topics to be covered. One of the requirements was that not only theory but also practice should be taught.

'Map Your World' is a project created in 2009 at the Institute of Geomatics Engineering (IGE) at the FHNW. It is basically a project to familiarize secondary school students with both modern geo-sensor technologies such as global navigation satellite systems (GNSS) and concepts of contemporary cartography and web-mapping applications, such as Google Maps. In addition to these basic intentions, the project aimed to reveal the complex and, with all of its applied technologies, quite often very demanding task of making a map (Stark & Treuthardt, 2012). The drawback to 'Map Your World' was that it took much more than four lessons to be taught. Hence a new approach had to be found to meet the four-hour limit without losing too many of the pedagogical benefits of 'Map Your World'.

GPS tracks of the students' fieldwork existed from earlier 'Map Your World' projects, and a new approach of analysing these same tracks was developed. Henceforth, the focus within the initiative was on the analysis of the students' patterns of movement and their leisure behaviour. This new project was named 'See You'. This name refers, on the one hand, to the leisure time in which the tracks are logged, and on the other to the analysis of the collected track data which is to be investigated, to be *looked at*. For the 'See You' project, secondary school students carry a GPS logger for a specified time, for example a week or a few days. They are asked, especially in their leisure time, to have the GPS logger with them to record the places they go to, and also to document the routes they take. As a result, the tracks of a class of secondary students within a given perimeter are created and are then available for further investigation. Of special interest are the popular places they go too often. In this project *popular places* include not only indoor but also outdoor locations.

2 Description of the Learning Environment

The Web Application. A web-based application was developed which can be used both to upload GPS tracks from the students and to analyse the collected data. The application is password-protected and requires valid credentials which the teacher delivers to the students. This security is necessary to guarantee the privacy of the participating students.

The main web page shows the map in the centre and, on the left, the different map layers to be used (see Fig. 13-1) and the option to display extracted indoor locations and hotspots. There is a choice of four different background layers. The resulting density map of hotspots along with the hotspot layer can be switched on or off for display.

Figure 13-1: Main webpage with density map, hotspots and applied filters

Indoor locations indicate buildings that the students entered and where they stayed for a certain time. There are four categories of indoor location: the first are places with an indoor stay of between 15 and 45 minutes, indicating potential shopping places. The second category is for indoor places with a stay of between 45 minutes and 3 hours, indicating potential leisure places like cinemas, restaurants or sports venues. The third category is for indoor places with a stay of between 3 and 8 hours, indicating places like school. Finally, the fourth category indicates places of residence, with a stay of between 8 and 24 hours.

The application provides a layer of hotspots that indicate the pixel areas with the highest pixel values, i.e. places where many track points were recorded. This hotspot layer helps to find the places that students visited the most or where they stayed the longest. Hotspots are presented in the map with a circle and an index number. The number indicates the ranking: the lower the number, the higher the ranking, i.e. the more track points were recorded. The number of track points is determined either by the duration of a stay – the longer a user stays in one place, the more track points are recorded at that location – or by the recurrence of a visit. Hence, hotspots are subdivided into two categories: hotspots according to duration of stay, and hotspots according to frequency or recurrence of visit. The former are classified with a ranking number where 1 is the highest ranking; for the latter, the absolute number of visits is recorded.

Below the map, there are a few filter options which can be applied. It is possible to analyse only GPS data which was collected within a specific time window ('from … to …' values), or from students within a specific age range (slider) or of a specific gender. The gender is specified by checkboxes.

When all the settings are chosen, pressing the calculate button at the bottom will compute the density map, which then appears within a few seconds in the map window. There is a slider on the left which allows for a reduction or an increase in the number of hotspots to be shown on the map, or to regulate the transparency of the density map.

In addition to the spatial dimension, the temporal component is also of interest. The GPS tracks contain the time stamps of the registered positions. Based on this information, the application also provides video sequences of the density maps. These video sequences can be downloaded, studied individually or in class, and discussed. The video may indicate specific time windows when students are more on the move than at others.

Self-declarations of popular places by students are also possible by placing points of interest (POIs) on the map, to which comments can be added by other students. This basically 'a-priori' definition of specific leisure places can then be compared to the analyses of the GPS tracks collected and may lead to an interesting discussion of perception and fact.

Data Processing. The base input data set consists of raw GPS tracks which can come from any kind of GPS-enabled sensor. In the 'See You' project, low-cost GPS loggers were used (see Fig. 13-2). These loggers have the capacity to store the tracked positions as values of longitude and latitude in the World Geodetic System, which can be read out later. Another great advantage of these devices is that they are lightweight and smaller than a pen, and thus fit easily into a pocket. The Navilock NL-457DL EasyLogger contains a small software package which allows for the signal to be logged. The configuration settings of the Navilock NL-457DL EasyLogger were optimized to obtain the best possible tracking results. The i-GotU GT 120 logger (see Fig. 2) requires some additional software to download the collected data either to a Windows-based PC or to a Mac OS-based Apple computer. Any GPS sensor can be used for the collection of the tracks, be it a smartphone or a more sophisticated GPS device designed for outdoor activities.

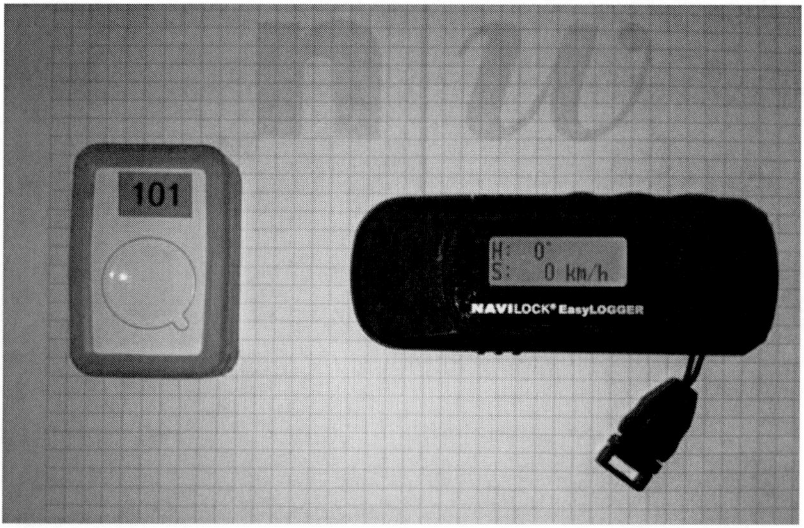

Figure 13-2: i-GotU GT 120 logger (left) and Navilock NL-457DL EasyLogger (right)

Since the GPS signal is not good at all times and under all circumstances, the computation of position is sometimes error-prone. This is either because too few satellites are available, or because the satellite signals are disrupted and a good fix on the current position is impossible. To avoid processing GPS tracks of bad quality which may bias the results, some post-processing of the raw GPS data is performed before the spatial pattern analysis is carried out. The main goal is to detect and erase outlying positions. The filters used are, on the one hand, large time slots between consecutive point registrations and, on the other hand, segments where movement was above the city centre speed limit.

Another issue that has a negative impact on the tracks recorded is the initialization process, which lasts until the device receives a clear and unambiguous GNSS signal. While the correct position is being detected, the device is already logging positions. This leads to a 'jumping pattern' at the locations where the devices are switched on (see Fig. 13-3). How to improve the quality of the tracks and erase these anomalies is currently being researched at IGE.

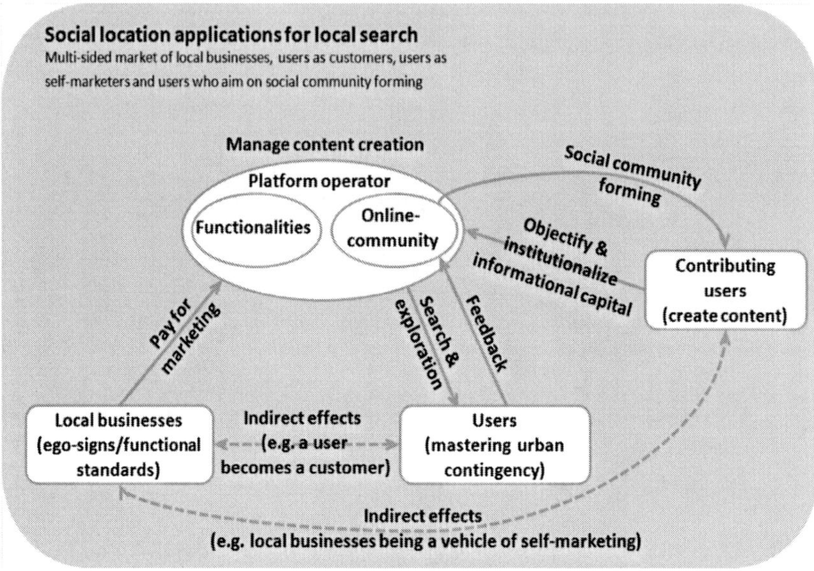

Figure 13-3: Jumping Tracks visualized in Quantum-GIS with backdrop map data from OpenStreetMap.

Analysing Spatial Patterns of Leisure Behaviour. Spatial patterns of leisure behaviour on a small scale have been investigated for quite some time and have enjoyed good support since the widespread use of GNSS. A summary of the findings and research is presented in Stark (2012).

A density map as mentioned above is the base upon which the leisure behaviour of secondary students is analysed within the 'See You' project. After the cleaning of the raw GPS tracks, the data from all the students is uploaded into a central database via the Internet. The extent of the spatial dimension of the study area can be defined at the creation of a new project, or it is automatically computed based on the limits of all tracks; the limits are also stored in the database. Based on the area covered and the dimension of the resulting density map, the size of a grid cell is computed. 'See You' uses an absolute counting approach for each grid cell. Since there is always an element of inaccuracy in GPS tracks, it may be misleading to take into consideration only those grid cells which are touched by a GPS track segment. The project's approach allows for the application of a degree of tolerance by using a radius r parameter. The basic concept is as follows: for each GPS segment, the minimum bounding box, which is extended by the given radius r, is computed. Within this bounding box, for each pixel the distance to the segment line is computed and, based on the inverse distance, an estimator of proximity is computed. If the pixel falls outside the buffer which is defined around the line with the distance r, the pixel receives a null value. Figure 13-4 illustrates this algorithm. The computed values are dimensionless.

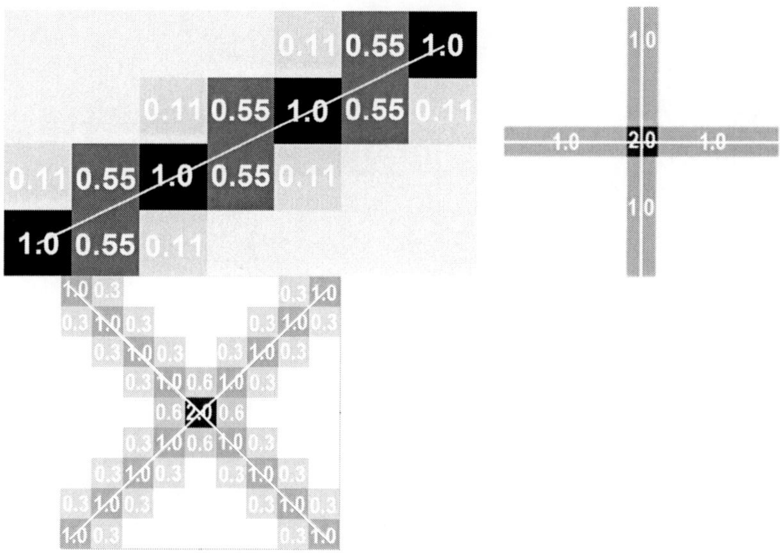

Figure 13-4: Computation of grid cell values for different segment lines with given radius r = 1. The image on the left shows a diagonal line and the computed grid cell values. The images below and on the right show the computed grid cell values for two segments of a different position and orientation that cross each other.

This procedure is repeated for each segment in all of the GPS tracks collected. The cell values are added up and result in the density map. Figure 13-5 shows an excerpt of the density map of some sample tracks in the city of Basel, Switzerland.

Figure 13-5: Density map for a set of GPS tracks in Basel without backdrop map data (left), slightly zoomed in, and with backdrop map data from OpenStreetMap (right).

3 Preparation for the Teacher

Setting up a new Project. Setting up a new 'See You' project is very easy. The complete infrastructure, from the GPS loggers to the ready-to-use application and teaching materials, is available and provided free of charge by IGE. The only prerequisite is a computer with an Internet connection and a standard web browser. First of all, a teacher sends an e-mail to *info@see-you.ch* and requests the necessary details to start a project. The 'See You' project management sends both a username and a password back. In a standard web browser, the teacher goes to the *http://app.see-you.ch* website and logs in using the username and password supplied. After that, the teacher creates a project as described below.

To run the application, the teacher has to set up a new project (see Fig. 13-6, left). Creating a name and password for the project, the teacher can define the perimeter of the area of interest. After opening the project, the students are able to upload their GPS tracks directly from the logger via a standard web browser (see Fig. 13-6, right). No additional software plug-in is necessary. In addition to the tracks, the students are asked to enter their gender and age. This is additional information which can be used for further analysis (filtering). Once all the data is uploaded into the central database, the application is ready for analysis to be performed by both the students and the teacher.

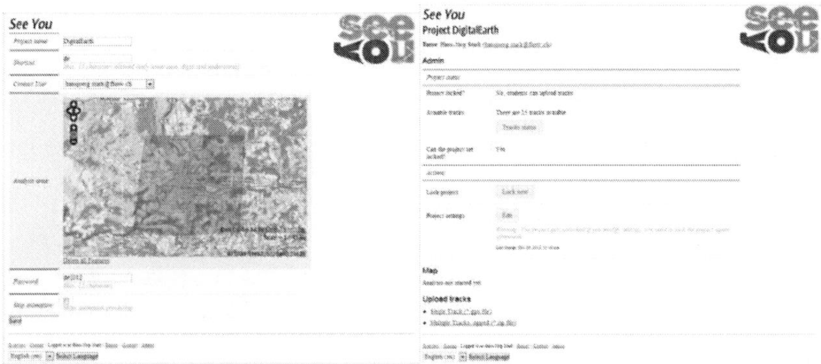

Figure 13-6: Web interfaces for project opening and track upload

Process and Duration. It is recommended that the teacher uses the first lesson to introduce the project in the classroom. They may want to explain how GNSS works and how to navigate the application. The available project demo serves as a first entry point: it provides data from a single user, and by applying different filters or displaying hotspots and indoor locations, the students become familiar both with the functionality and the meaningfulness of the data presented. Simple questions like 'Where does the person who recorded the data live?', 'Where does he/she work?', 'What are his/her favourite places?', 'Which routes does he/she often use?', 'Are there significant nodes of everyday action to be detected?' etc. help the students focus on the (visual) analysis of the map.

In the second lesson, the students go outside and familiarize themselves with the GNSS loggers. They do some tracking and upload the data into a sample project set up by the teacher for the purposes of learning the upload and post-processing procedures.

These two introductory lessons are followed by the period given over to recording movement. We would recommend that the students record their movements for a week.

The third lesson concerns uploading the collected data and analysing it within the web application. The students are asked to use their imagination in this analysis process and to note their findings (cf. Fig. 13-7).

The fourth lesson finalizes both the experience with the logger and the findings. The students are asked to discuss their research results with their peers while the teacher acts as the presenter. He/she specifically draws the students' attention to issues of privacy and their potential danger (cf. Wang & Loui 2009). Last but not least, one of the main emphases of this project is to make students aware that tracking with smartphones and similar devices can be performed without the knowledge of the person being tracked.

Figure 13-7: Students exploring their 'See You' project

The privacy issue is stressed through the analysis of relations between tracks that 'See You' provides: the project analyses the GPS tracks based on the information of who the recording person was, and the track information of location (co-ordinates) and time. This means that, when two tracks are very close and the recorded locations are registered at nearly the same time, it can be assumed that the people who were carrying the GPS loggers have a closer relation to one another. The time-duration of this matching constellation of GPS tracks and the frequency indicate the strength of the relationship. The result is a matrix which shows for each participant how much time they spent with their friends (see Fig. 13-8). It must be emphasized, however, that the figures in the relation analysis are not to be taken literally, because their determination depends very strongly on the handling reliability of the person who was carrying the sensor. We need to consider, for example, whether they switched the sensor on and off on time, whether they waited for good GPS signals, etc. Nonetheless, the relation analysis gives an impression or at least an indication of the different levels and strengths of the relationships between the students.

Relation-Analysis www.see--you.ch

	martin	lukas	enzo	philipp	markus	sarah	martina	pascal	patrick	raphael	michael	silvan	thomas	michelle	fadri	pascal	jeannette
martin	-																
lukas		-															2
enzo			-						5	4		1			20		
philipp				-	9								1	12	2		
markus				9	-									13			
sarah						-	29										
martina						29	-										
pascal								-			83						
patrick			5						-	33		26			1		
raphael			4						33	-		92					
michael								83			-						
silvan			1						26	92		-					
thomas				1									-				
michelle				12	13									-			
fadri			20	2					1						-		
pascal																-	
jeannette		2															-

unit [min]

Table 13-1: GPS track-based relation analysis

Teaching Material. At the time of writing, application to the project is available at *app.see-you.ch*. General project information is available at *www.see-you.ch*. Both are available in English and German, as is the teaching material.

Student Assessment. As mentioned earlier, the fourth lesson finalizes the findings. The students managed to become familiar with the application in a very short time, and the closing discussions within class on the findings of the students based on their analysis were very interesting. Some of the students' results were:

> 'On Tuesday afternoon, someone was at the dentist's.'
>
> 'When we're out, the boys are together and the girls are together, except on Friday afternoon.'
>
> 'There is no pedestrian crossing in front of the large shopping centre, even though many people cross the street there.'
>
> 'You could figure out where to place a kebab stand.'

'On Sunday afternoon, Peter did a bike tour in the forest. You know this because there's a red point in front of Peter's house and the route he took in that time is too far to jog or walk.'

'The boys were out longer than the girls, but the girls were further away.'

4 Conclusion

The project developed an easy-to-use and widely accessible web application which allows for both the teacher and the students to analyse and discuss the spatial patterns of their leisure behaviour. The time constraint of only four lessons, while limited, is manageable. A great advantage of the approach presented is that the teachers do not have to invest a great deal of time in the preparation and learning of new software, like GIS or data editors. The technical skills required are within easy reach of most students and teachers. The GPS logger is very easy to use, and a standard web browser is all that is needed for navigating. There is the option to use the GPS loggers provided, but this is not a pre-requisite.

The 'See You' project was ready to use in spring 2012. Since then, it has been run several times in different settings. The basic idea of this application is that both the teacher and the students are able to easily navigate in the map of their study area. They can go through different analyses and discuss the computed density maps. For example, they may want to know where the hotspots on the weekends or in the afternoons between 14h and 18h are for the girls, etc. The maps should foster the students' awareness of their habits and the spatial patterns they leave. A special focus in the analysis in class is on what their tracks tell about their private lives. As the projects carried out so far have shown, these goals were reached.

Overall, the project was appreciated by both teachers and students and by people from outside the school environment.

It can be assumed that there is a difference when performing the project in winter or summer time. The students probably spend much more time outdoors in warmer months than they do in winter. No research or comparison has been conducted on this issue yet.

Further improvements to the project design will be made and initial 'best practices' will be elaborated based on the students' experiences. Since the individual projects are password-protected, no public view is possible.

The project was not only applied by teachers but also taken up by a social worker who used it to define, identify and discuss important leisure places for teenagers in a specific urban area. This information is of high value for other professionals in areas such as urban planning, and 'See You' has proved to be a valid and valuable tool for this task.

References

Stark, H.-J. & Treuthardt, C. (2012): 'Switzerland: Introducing Geo-Sensor Technologies and Cartographic Concepts Through the Map Your World Project'. In A. J. Milson, A. Demirci and J. J. Kerski (eds.), *International Perspectives on Teaching and Learning with GIS in Secondary Schools*. Netherlands: Springer. 255–262.

Stark, H. J. (2012): 'Spatial Pattern Analysis of Secondary School Students' Leisure Tracks'. In GI_Forum, *Geovizualisation, Society and Learning Salzburg*. Berlin/Offenbach: Herbert Wichmann Verlag. 336-344.

Wang, J. L. & Loui, M. C. (2009): 'Privacy and Ethical Issues in Location-Based Tracking Systems'. *Technology and Society*. ISTAS '09, IEEE International Symposium .

CHAPTER FOURTEEN

ARGUMENTATION AND DIGITAL GEOMEDIA: THE EXAMPLE OF THE GOOGLE EARTH LAYER 'CRISIS IN DARFUR'

DETLEF KANWISCHER AND INGA GRYL

Abstract

There are several reasons why students should be able to analyse geopolitical argumentation with digital geomedia. First, digital geomedia are becoming more and more present in our everyday lives; second, digital geomedia are linked in a remarkable manner to people's everyday lives as they connect information and the spaces of everyday action; third, they are especially powerful as they combine visual and textual processing, as learning theories and educational psychology prove. The praxis-tested example of this paper uses the Google Earth layer 'Crisis in Darfur' to show how deconstruction applied as a practical method helps to identify hidden argumentation and to understand the processes of using geomedia. The paper provides ideas for transferring this highly political example to a broad variety of other applications in order to facilitate students' understanding of digital geomedia not only as social constructions, but also as instruments for arguing for certain interests.

1 Pedagogical Background, Aims and Target Groups

Millions of young people 'fly' over continents, mountains and cities, familiarize themselves with their next vacation spot, or measure their way to school: digital maps and globes have established themselves firmly in our own and our students' everyday lives over the last decade. Due to their simple user interface, they enable us to create maps which are interactive and in which a variety of spatial information in the form of images, texts, statistics, etc. can be embedded. These digital maps and globes give the map consumer the possibility to simultaneously contribute to the whole production process, as a so-called map prosumer.

These developments offer possibilities to use digital maps and globes in the classroom in a versatile way, which is especially useful as many problematic issues dealt with in geography classes are presented in a digital and geo-referential way. For example, Google Earth provides a small number of layers which can be used to direct attention to global problems ('Global Learning'). The Google Earth layer of the project 'Crisis in Darfur' has been developed in cooperation with the United States Holocaust Memorial Museum (USHMM). In this, the area of the conflict is linked to multi-media and geo-referenced data (see Fig. 14-1).

This example suggests that geo-referential argumentation procedures will be increasingly applied in the future. However, the arguments underlying the objectives of this information should be questioned, and consideration should be given to which counter-arguments might need to be developed. In order to participate in this discourse, other arguments have to be understood and questioned, and one's own argumentation has to be formulated – which, ideally, will also be digitally visualized to communicate geo-referenced data effectively.

Using the layer 'Crisis in Darfur' as a classroom example in order to further students' argumentation competences with the help of geomedia requires the students to have a certain understanding of political and historical relations. Therefore, the example is aimed at a target group of students of 14 to 18 years old.

Theoretical Background: Geomedia as Social Constructions and Instruments of Argumentation. 'The Power of Maps' (Wood, 1993) is the title of a classic publication within critical cartography that established a keyword which accentuates the potency of argumentation that uses maps. However, it is not the medium itself that is powerful here; rather, the map serves as a means to convincingly distribute and implement spatial world images and ideas. Maps and any other geomedia are particularly effective since they possess an inherent credibility (Crampton, 2001; Wood, 1993). One reason for this seems to be the positioning of information and its interpretation in the physical world. In this way, cartographically communicated interpretations are directly connected to a world which is perceived as real.

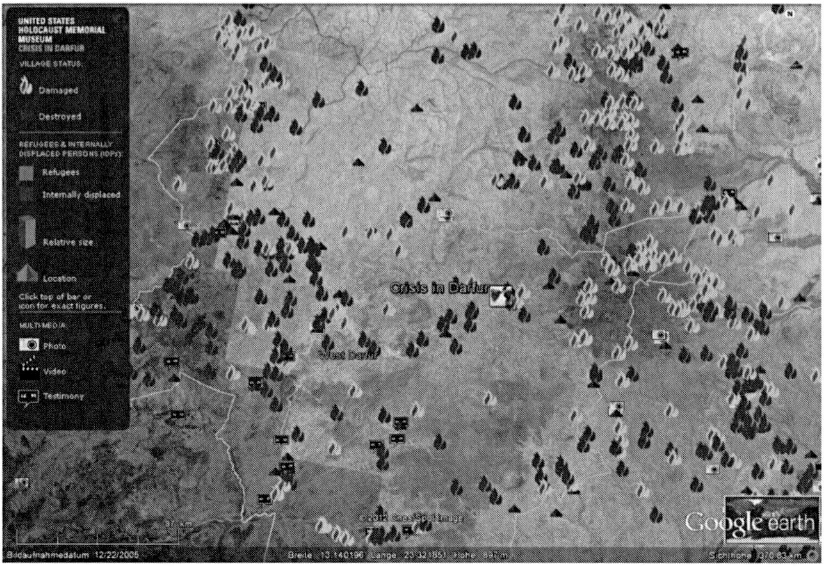

Figure 14-1: Interactive presentation of the area of the Darfur conflict on Google Earth (Screenshot of Google Earth)

Indeed, orientation with and within geomedia requires knowing that the geomedium focuses on a specific perspective, as it offers access not simply to information, but to information which is reduced in order to make it intelligible. Within the process of translating aspects of the world into cartographic signs, selection and standardization take place. Furthermore, the usefulness of geomedia for the consumer is defined by adapting the information presented to the daily and common spatial problems that the consumer is confronted with. Subsequently, the author (or authors) of the geomedium decides consciously and unconsciously which contents will be included and how these will be displayed. Providing enormous scope for subjectivity, the geomedium is converted into a medium for argumentation itself, or as a complement to verbal and written argumentation.

The following classroom example using the 'Crisis in Darfur' layer will illustrate that the repercussions generated by the argumentation of geomedia depend on the consumer, who no longer perceives a one-to-one relationship of the medium with the 'real' world. In fact, the consumer – consciously or unconsciously – uses the medium as a starting point of thought processes in order to construct his or her own world view (Crampton, 2001, 245). Self-directing this process is a starting point to maturity in working with geomedia. Deconstructive analysis of argumentation structures will support this ability, and will constitute the basis for an active digital geomedia production which is needed for developing argumentation competence.

If one identifies argumentation competence as a competence with three elements (reception, interaction and production) (Budke, Schiefele & Uhlenwinkel, 2010, 186), then two main perspectives emerge regarding argumentation with geomedia.

Pedagogical Background I: Reflexive Geomedia Competence for a Mature Reception of Argumentation in/with Geomedia. In order to responsibly appropriate space, that is to say to consciously assign meaning to things and thereby to direct one's own actions purposefully and in harmony with the animate and inanimate environment, geomedia have to be read in a (self-)reflexive way. Notably, critical cartography points out that what is not represented in a geomedium, and thus exists outside the discourse of the geomedium, deserves special attention. These aspects lead towards the concept of reflexive geomedia competence within the framework of geography education. Starting with context-free knowledge about the constructedness of geomedia by focusing on different constructional elements, which the map author selects, this type of knowledge will be presented and then applied by working with a geomedium. The change of perspective towards things outside the geomedium's discourse will be assured by comparing the geomedium with external information. Evaluation of the geomedium will not be according to categories of 'right' or 'wrong'; rather, the assessment of the geomedium's suitability to represent a spatial problem will be made against the background of the limitations of its discourse. Taking the idea of geomedia as instruments of argumentation, Budke et al.'s approach of argumentation competence can be transferred to geomedia: with regard to the issue of decoding, students on the lowest competence level recognize 'opinions' as such (Budke, Schiefele & Uhlenwinkel, 2010, 186); on a higher competence level, they are increasingly able to assess more complex arguments.

Competence related to geomedia – and, especially, reflexive geomedia competence – means that students can reflect on cartographical argumentation, evident in every geomedium's construction of the world, concerning its appropriateness and explanatory value for specific problems. It is precisely this reflective assessment that allows one to review the geomedium's information in terms of its value for further usage, or to identify the need for additional material.

Pedagogical Background II: Active Geomedia Production as a Means for Argumentation. According to Budke et al.'s model of stages (2010, 186) and within the framework of competence-development related to the production of arguments, students first voice their opinion about a geographical issue. Increasing competence will then lead to a justification of this opinion with appropriate and valid argumentation; the justification will also be tailored to its intended audience and display multiple perspectives.

A sign of effective argumentation is the ability to look at problems from multiple perspectives (Budke, Schiefele & Uhlenwinkel, 2010, 186); however, the discourse of a single argumentation is always limited, and so is every cartographical argumentation. Discourse concerning the different interpretations inherent within cartographical argumentation first emerged when the one-sidedness of cartographical production, which itself has always been an elaborate and expensive process limited to specific institutions and companies, was reduced and cartographical production became an everyday activity for many. Therefore, the abilities and skills to produce geomedia have to be understood as a communication skill, as speaking a language that voices one's own opinion in a comprehensible way, while at the same time using geomedia to support a convincing argumentation. Since the emergence of more and more user-friendly tools for cartography, in particular on the Internet, which offer free data sets (or free for non-commercial use), the formerly expensive computer cartography has arrived in contemporary everyday life, at least in its simple, basic forms. Today, active lay cartographers exist. Even though it is not possible to use these tools for spatial analysis in the same way as professional GIS applications allow, the visual and graphic results are comparable to the results of professional cartography, even with only a basic technical competence. Students can thus communicate their argumentations effectively to their specific audiences. In addition, students can be made aware of the multiple perspectives of argumentation by juxtaposing the layers of a digital map's different interpretations. At the same time, geomedia are not one-sided argumentation tools, but a means for oral and written argumentation and, consequently, a vehicle for the interaction of responsible special interest groups within social negotiation processes; interaction is regarded as a further dimension of argumentation competence, as Budke at al. point out (2010, 184). Thus, in order to promote argumentation competence, a reflexive interaction with geomedia is particularly effective.

2 Description of the Learning Environment

The qualities of geomedia as a medium of argumentation – as described above – have consequences for working with geomedia within geography education. Geomedia, and especially maps, still exist as a medium for conveying content knowledge, although to understand and continue working with geomedia in the classical sense of reading and interpreting them no longer suffice. Analysing geomedia has to go hand in hand with assessing them; producing geomedia has to be accompanied by reflecting on one's own geomedia production.

The following steps can help to develop argumentation competence by means of classroom work using digital globes and maps.

Step 1. Students conduct an online literature research about Darfur and collect facts from different sources, thereby gaining an overview of the Darfur crisis. The material provided in section 3 below, especially M1 and M2, offers valuable summaries of the conflict. The students then investigate further sources which may confirm the reports given in M1 and M2 and can add new information. In doing so, students will be sensitized to recognizing which sources provide which type of information. The information collected can be visualized in a graphical concept map.

Step 2. It is now important to analyse how the conflict has been understood by the students up to this point. On the basis of the completed concept maps, the conflict is further characterized and illustrated, with all its existing interrelations. M3 explains some conflict types with regards to the Darfur crisis, which are then compared to the current information on the concept map and discussed in the classroom.

Step 3. After the students have gained a general view of the conflict and discussed the characterization of the conflict, the website of the USHMM mapping initiative (http://www.ushmm.org/maps) can be opened. 'Crisis in Darfur' was one of the first projects that the USHMM launched in cooperation with Google's initiative 'Genocide Prevention Mapping Initiative'. On the first page, the project is introduced and the technical requirements needed to install the layer 'Crisis in Darfur' into Google Earth are explained. Google Earth serves as a medial intersection for a collection of many further geo-referential media representations in the form of maps, photography, statistics, reports and interview statements from the people involved. The project's aim is to convert the traditionally long and investigative reports of the UN and other organizations into a format that can set in motion a public discussion and, consequently, an informed intervention:

> 'We at Google are convinced that technology can be a catalyst for education and action.' (Google 2007, according to Parks, 2009, 431)

Step 4. In further classroom work, it is important for the students to understand how the conflict has been represented, and to question the arguments of the authors of the layer and of other illustrations and texts. Budke and Uhlenwinkel (2011, 3–4, abridged, translated and amended slightly) provide six aspects useful in the field of argumentation competence and especially the competences of analysis of argumentation:

- Analysis of 'completeness' (assessment of gaps visible when comparing different sources, while being aware of the incompleteness of all information sources)
- Analysis of 'complexity' (analysis looking for approaches to multi-perspectivity through counter-arguments)
- Analysis of 'relevance' (assessment of relevance of the information/arguments for the problem)
- Analysis of 'validity' (assessment of validity – as far as can be appraised at all – of sources by taking into account their context of production, while being aware of the general social constructedness of information)
- Analysis of 'eligibility' (assessment of internal consistency and stringency between sources and statements, and their cogency)
- Assessment of 'recipient-orientation' (assessment of the sources' orientation to recipients in terms of influencing and convincing the recipient, or encouraging the recipient to take responsibility for his/her thinking processes).

The following key questions serve as a starting point for analysis in relation to the Darfur case, and can be used in geomedia analysis in general:

- What is the structure of the line of argumentation? (completeness and complexity analysis)
- What is missing in the argumentation in comparison to other sources? (completeness and complexity analysis)
- Which information is used, and is this relevant for the statements as well as for the problem described? (analysis of relevance)

- Which opinions/claims are being voiced, and which sources are given? (analysis of validity)
- Are the proposed interrelations demonstrated consistently and stringently? (analysis of eligibility)
- Which recipients are being addressed, in what way and with what aim? (recipient-orientation)

Discussing these questions will lead to the students' realization that the history of postcolonial geo-politics in the Darfur region has not been taken into account within the Google Earth layer; information about the different peace treaties is missing; details about development aid workers and peace troops who were killed have not been mentioned, and information about perpetrators of violence, their resources and military movements is scarce (Parks, 2009). To conclude this stage, students might reflect on how the requirements of the layer production are connected to the illustration of the layers. In other words: Who speaks from which point of view, about what, and for whom, with what purpose? It becomes obvious that behind the message about aid a PR-message is hidden, eventually aiming to present Google as a humanitarian company.

Step 5. Following this deconstruction of layer production, the final task could be to produce one's own layer with a more complete structure of argumentation. Information to be included has been identified, and is provided, in steps 1 and 2. Creating one's own layer can be done relatively easily (see http://www.ushmm.org/maps/projects/darfur/ → Create your own presentation).

3 Preparation for the Teacher

In the face of complex problems, the involvement of many different groups and the lack of an internal cultural perspective, conflicts and civil wars in Africa are often difficult to explain. What can we say about this in schools? How can we understand these conflicts by applying our own categories and worldviews to them? It helps little if the civil war in Sudan is explained with categories such as (Arabic) Islam in the North against (black) Christianity in the South. The students' competence first needs to be developed methodically in order to enable them to identify existing argumentation patterns around the conflict and to justify their own argumentation. The following material may provide additional background content, delivering basic information on the most important relationships in the conflict, and providing an additional perspective to the Google Earth layer.

M1 A Short History of South Sudan

The referendum in South Sudan that took place 9–16 January 2011 marks a historical turning point – not only for a country in which a civil war lasted/raged for half a century. […]
The hostility between the North and South reaches back to the times when the black Africans of the South were hunted by the Arab slave traders from the North. After Great Britain ordered a union with the North, many people in the South converted to Christianity in order to demonstrate their non-Islamic identity. The British did nothing to overcome the deep separation within their colony; they governed the North and the South as virtually separate regions. In doing so they favoured the North by investing in its social and economic infrastructure. The Civil War, which had begun before the declaration of independence, did not end until 1972. […]
New tensions were triggered when the US oil company Chevron extended its activities in South Sudan, where 85% of the Sudan's oil reserves discovered so far exists. The former President Dschafer al-Numeiri attempted to abolish the hard-fought autonomy of the South. He closed the parliament in Juba, disposed of the regional government, and installed a military administration. When the military administration tried to disband the black African regiment, the soldiers rebelled. This incident caused the second inner Sudanese war in May 1983, which lasted 19 years. […] Under the government of George W. Bush, Washington again intervened in the conflict. Pressured by the religious right who made the issue of the 'persecuted Christians in South Sudan' into their own, the Bush administration brought the hostile parties to the negotiating table in October 2002. Three years later a peace treaty was signed in Nairobi. The Comprehensive Peace Agreement (CPA) provided for a referendum about the independence of the South to be held within the next six-and-a-half years. […]
The future remains uncertain. Certainly the old problems will appear again, foremost the question of the ultimate borders and the distribution of oil profits. Nobody can say what will happen in the North. However, Khartoum remains nervous, and this makes the situation insecure if not outright explosive.

Material 1, source: abridged and translated, Prunier, G. (2011): 'Kleine Geschichte des Südsudan'. *Le Monde diplomatique*, no. 9418, 2.11.2011, 10–11.

M2 South Sudan Celebrates its Independence

South Sudan has officially declared its independence. The president of the parliament, James Wani Igga, read the declaration of independence at a ceremony in the capital, Juba. Cheered by a crowd of thousands of people, the Sudanese flag was lowered and the flag of South Sudan was raised. Subsequently, the new President, Salva Kiir, signed the Transitional Constitution. The Sudanese President Omar al-Baschir, who is wanted for war crimes in Darfur and against whom an international arrest warrant has been issued, attended the independence celebration alongside heads of state, high officials and dignitaries from all over the world. North Sudan had officially acknowledged the independence of the South at the last moment, on Friday. Germany also immediately entered into diplomatic relations with South Sudan.

Material 2, source: abridged and translated, Moutafis, G. (2011): 'Der Südsudan feiert seine Unabhängigkeit'. *Zeit Online*, 9.7.2011 <http://www.zeit.de/politik/ausland/2011–07/suedsudan-staatsgruendung-unabhaengigkeit> (accessed: 17.7.2011).

However, the German World Hunger Organization (Agro Action, Hunger Relief) warned of the danger of civil war in South Sudan:

> 'Thinking that the separation of the Sudan will pass without conflicts would be very careless,' says the president of the World Hunger Organization, Bärbel Diekmann. The situation in the South Sudan seems very tense. 'Our staff report emaciated people, some of whom have begun to eat seed.' This will lead to failures in the next crop. Approximately 40% of all children in South Sudan are undernourished.

The different types of conflict overlap and the dynamics of a locally rooted conflict have changed over time; nor can the different phases of this conflict be distinguished exactly from one another.

The basic technical requirements for this project are just one teacher's computer connected to a Smartboard and a projector. However, letting the students explore the layer on their own will significantly increase their insight and competence regarding the analysis of geomedia argumentation. For Internet research on the Darfur crisis and in order to construct their own maps with Google Earth, students need web-connected computers, laptops or tablets. Google Earth can be installed for free, but administrator rights on the target computers are required. For collecting the differing student-produced layers and visualizing them in front of the class, a common cloud or server to store the results is useful.

M3 Local Causes, Global Impact

The conflict in Darfur remains complex; different factors come together at local, regional, national and international levels. The conflict is constantly reignited because the different parties seem to be locked in a spiral of violence. In this region, the conflict is no long just between black Africans and Arabs […].
The region is populated by more than 100 tribes and clans. Conflicts within a group are as frequent as between different groups. Armed robbery and theft are an everyday occurrence and have devastating consequences because of access to modern weaponry. I will call this conflict type 'type A'.
'Type B' conflict has regional dimensions. Two collective entities which are not strictly separated from each other – the Zurgha (black Africans) and the Arabs – compete for political resources.
At the heart of this conflict lies the question of social, collective and even racial identity, although it is also caused by tensions of type-A conflicts (over land, water and pasture). This competition, expressing itself in the category of identity politics, escalated, and the emergence of Islamic Fundamentalism in Central Sudan has greatly contributed to it. More and more, type-C conflicts overshadow the identity-based type-B conflicts: i.e. rivalry about power, resources and prosperity between the total population of Darfur and Central Sudan.
Type D, on the other hand, encompasses the international dimension of the conflict in Darfur. Other African states and the most powerful nations in the world observe what is happening in Darfur. On this level, the conflict is not only about legal principles and the cost of this humanitarian crisis, but also about access to sources of energy. While most African governments tend to support the government in Khartoum, the G7 has expressed its criticism of it.

Material 3 (accessed: 12.10.2012), source: abridged and translated, El-Battahani, E. (2006): 'Lokale Ursachen, globale Folgen'. In: *Entwicklung und Zusammenarbeit*, <http:/www.inwent.org/E+Z/content/archiv-ger/11-2006/trip_art3.html>

4 Conclusion: Evaluation and Transferability

The classroom example illustrates the opportunities that digital maps and digital globes possess for the development both of argumentation competence and of (reflexive) geomedia competence. It is predicted that these competences will gain in importance since digital maps and globes will become central and universal communication interfaces of spatial information in the near future.

The example has been shown to work with students aged 14 upwards within more or less traditional classroom situations. The limited nature of the technical requirements contributes to 'learning *with* geomedia'. The students gained an idea of the limitations of argumentation within the given geomedium, the complexity of the conflict, and the opportunities to produce counter maps with a more complex argumentation structure in a technically user-friendly way. It is important to encourage the students' creativity in producing these maps, and their confidence to provide valuable insights for others by questioning existing 'official' geomedia and their explicit and hidden intentions.

Argumentation competence regarding maps and other geomedia may be furthered in students even younger if appropriate examples, closely connected to their everyday lives, are chosen. The field of application for work on geomedia argumentation is broad, ranging from assessing global geographical problems to working on contested fields in the students' local area, such as aspects of town planning

References

Budke, A., Schiefele, U. & Uhlenwinkel, A. (2010): 'Entwicklung eines Argumentationskompetenzmodells für den Geographieunterricht.' *Guid* 3, 180–190.

Budke, A. & Uhlenwinkel, A. (2011): 'Argumentieren im Geographieunterricht – Theoretische Grundlagen und unterrichtspraktische Umsetzungen'. In C. Meyer, R. Henry & G. Stöber (eds.), *Geographische Bildung. Kompetenzen in der didaktischen Forschung und Schulpraxis.* Tagungsband zum HGD-Symposium. Braunschweig: Westermann.

Crampton, J. W. (2001): 'Maps as Social Constructions. Power, Communication, and Visualization'. *Progress in Human Geography* 25 (2), 235–252.

Parks, L. (2009): 'Ausgrabungen in Google Earth. Eine Analyse der "Darfur-Krise".' In J. Döring & T. Thielmann (eds.), *Mediengeographie. Theorie – Analyse – Diskussion.* Bielefeld. 431–453.

Wood, D. (1993): *The Power of Maps.* London: Guilford Publications.

CHAPTER FIFTEEN

USING A VISUAL ANALYTICS APPROACH FOR HYPOTHESIS-GENERATION

THOMAS JEKEL

Abstract

Recent research on Visual Analytics and critical GIScience suggests that diagrams, maps and spatial representations should not be seen as the end result of a research process, but as the starting point for enquiry (Crampton, 2001). Spatial representation, therefore, becomes a support tool for hypothesis-generation. This perspective allows a link to be made between visual analytics and a constructivist approach to learning and teaching with specific regard to science education. Students, through the learning environments presented, should get a feeling for the process of spatially-enabled hypothesis-generation as the basis for very elementary scientific research. In addition, visualization supports data-based argumentation in a variety of topics.

This contribution suggests two exemplary learning environments which need only one to two lessons to complete. The first is aimed at making the construction of everyday hypotheses from diagrams and spatial representations comprehensible; the second supports students with reasoning about phenomena in space using their own visualizations. The learning environments are both based on the free and openly available Gapminder software (www.gapminder.org), and have been successfully tested within 45-minute time slots with several classes of 16-year-olds at a GIS day event, making them eminently usable within limited timeframes and with minimal technical requirements.

Thematically, both learning environments are situated in the field of demography on global and nation-based scales. However, with the data available, the general approach can be easily transferred to topics such as economic development and crises, environmental issues or climate change indicators.

1 Aim, Target Group

Looking at European Union economies, one argument has been the hallmark for development in a competitive world economy, which is centred on the ability of school-leavers to innovate and produce completely original ideas. Innovation here is to be understood as the potential to bring together ideas or data which have not been connected earlier in a specific way. Innovation therefore asks for new interpretations and explanations of 'facts' or data – in short, to build causal hypotheses for real-world phenomena. It can be assumed that the development of new interpretations has so far not been at the core of secondary education, as these hypotheses are not easily classified as right or wrong by teachers. Furthermore, because of the openness of the concept, it may prove difficult to apply in standardized learning situations.

This situation may prove to be even truer if we consider images displaying data sets used in school books and teaching materials. Maps and diagrams have been designed to be read in a very specific way related to the message the author intends to convey to a wider public. The diagram or map here becomes the final act of a research endeavour, depicting the results of a process rather than the starting point. Current use of diagrams and maps can therefore, in many aspects, be linked to a linear communication model that transfers an author's or communicator's intentions directly to a recipient. As far as spatial information is concerned, the equivalent would be the map as a communication model, depicting the real world and communicating aspects of it to the audience (Crampton, 2001).

If the result of research is fixed within the image, two problems arise. First, if the result is shown in the image, no further innovation is necessary (the problem, so to say, has been solved). The image is there to provide a better impression and legibility of learning *content*. The technique, therefore, is to 'read' the image, extracting information already presented *with the aim of reproduction*. Second, and this relates to our sense-making of the world, this approach does not take into account that humans *construct* meaning from an image (or data) that may be rather different from the authors' intent.

What is suggested here, therefore, is the provision of learning environments that take into account both ideas: the individual and collective constructions of meaning from various data sets by using interactive and freely adjustable visualizations of data, as well as the basic argument for exploratory learning and own hypothesis-generation to support innovation in the classroom.

The target group here is secondary education, especially students aged 14–18, i.e. students whom we hope might become innovators within a few years.

2 Description of the Learning Environments

The cases presented here are both based on the freely and openly available Gapminder platform (www.gapminder.org). The first should enable students to reflect on their own abilities to build hypotheses, and deliberate on why their estimates and guesses have been 'wrong'; the second case is devoted to the free construction of hypotheses.

Gapminder includes tools and features allowing the depiction of up to five data sets in an interactive and dynamic image including time series. It helps its users to develop their own imagery from data, in a way that suits their argument but is still data-based. Extensive help with the site is available, including a 'gapminder for teachers' section. To explore hypothesis-generation and image-reception, all examples presented here use the chart function. The mapping tool, while visually pleasing, detracts from the tasks at hand, and seems more useful to reinforce very traditional world views.

Reflecting on Experience-Based Hypothesis-Building. The first case looks into experience-based estimates. To do this, students are asked to become population scientists, trying to provide an accurate estimate of the demographic development of Rwanda from 1985 onwards. Students were given some initial guidelines and presented with a short and very general introduction of demographic change. To support this, they were also given a depiction of life expectancy and birth rates for Austria in Gapminder (Fig. 15-1; available here at www.bit.ly/sBdWMI).

In the first step, students were asked for their interpretation of significant points on the graph for Austria. There was enough knowledge in the classroom to interpret significant jumps in fertility post-WWI due to returning soldiers, pre-WWII due to propaganda, and again in the 1960s due to economic development, as well as a sharp decrease in fertility due to new contraceptives in the late 1960s. At the same time, students could notice an overall decrease in fertility and an increase in life expectancy. It can therefore be postulated that students could read the diagram, as well as interpret it with the help of prior knowledge.

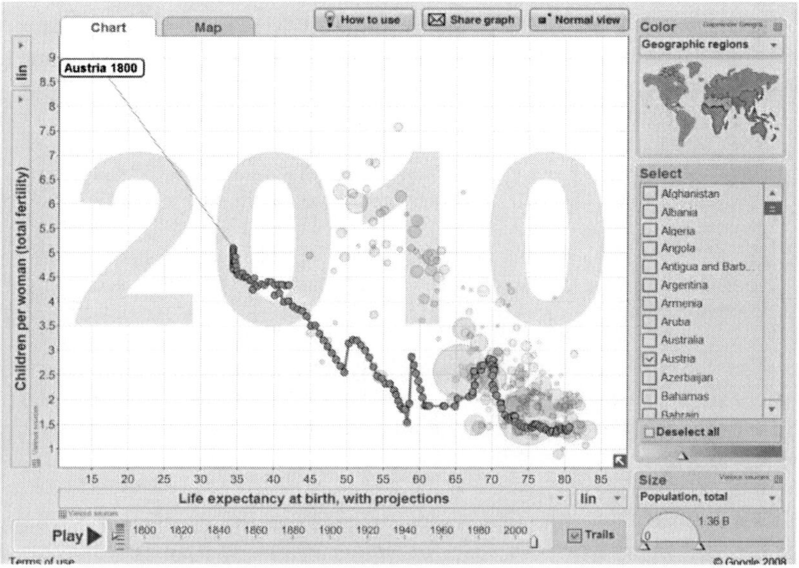

Figure 15-1: Fertility and Life Expectancy in Austria (www.bit.ly/sBdWMI)

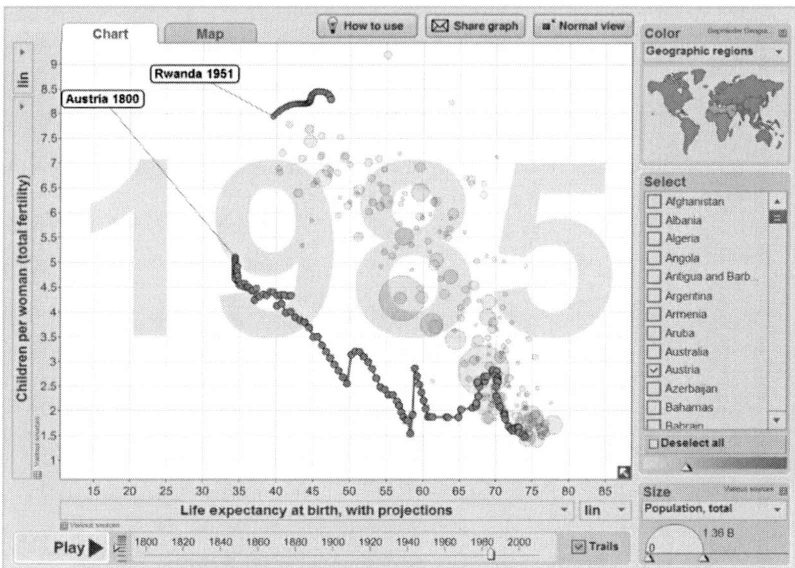

Figure 15-2: Basis for Fertility and Life Expectancy Estimates in Rwanda, 1985–2010 (www.bit.ly/142Nd0J)

In the second step, the graph was rolled back to 1985; in addition, Rwanda was selected (see Fig. 15-2). The image was projected onto a large piece of paper, and students were given the following task:

> 'Imagine yourself to be a population scientist in Rwanda in 1985 who is supposed to predict the development of fertility and life expectancy for Rwanda in the year 2010. You know that there are similarities in demographic change for most regions in the world, and you have seen the graph for Austria. Estimate carefully the development of Rwanda, write your initials on the coloured sticker and position it on the graph according to your estimate.'

The estimates of students would typically follow a fairly similar pattern to Austrian and global developments. Interestingly, none of the 16-year-olds in 2011 was aware of the genocide in Rwanda in the early 1990s, which led to a considerable shock once they were confronted with the 2010 data (see Fig. 15-3).

This shock gave rise to a series of questions that helped students to reflect on their everyday use of images and hypotheses based on seemingly 'safe' data. These included:

- What prior assumptions led to our wrong estimate?
- What type of additional data would we have needed to be more precise?
- Would a scientist have been able to give a more accurate estimate? (probably not)
- If this is the case, don't we have to be more careful with conclusions drawn in everyday life and science?
- What does that tell us about the process of making predictions about the future?

Depending on the specific classroom situation, any of these questions can be followed up and make up a lesson in its own right. These questions are, of course, transferable to other estimates of development, for example in the stock market. They allow the questioning of rules that seem to be stable, data that seem to be adequate for a specific question, and our own engagement with graphs as type of reasoning.

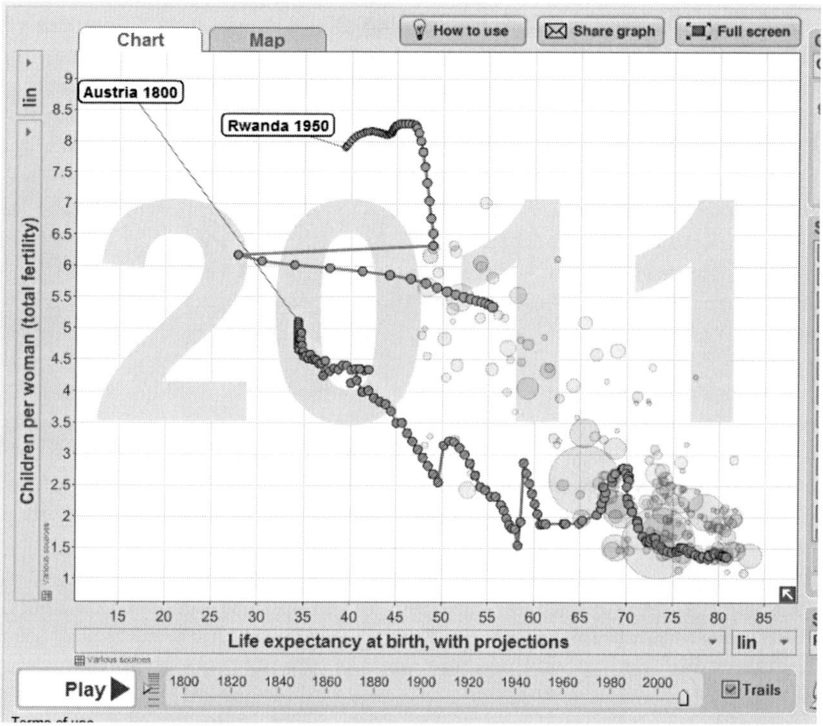

Figure 15-3: Development of Fertility and Life Expectancy for Rwanda and Austria up to 2011.

Building Hypotheses to Explain Specific Data Sets. While the first case presented here should make students aware of the basic mechanisms as well as the unreliability of predictions based on (spatial) data, the second is aimed at providing simple structures to organize problems, build hypotheses for explanation, and visualize for argumentation. For this exercise, let us return to the basic model of demographic change. One of the main questions here, of course, is: *What specific factors does fertility depend on?*

Students are then encouraged to use Gapminder as a very basic tool for visual analytics, looking for other indicators that seemingly explain the development of fertility rates over time. As most geographers know, a single explanation here is rather difficult to accept. The exploration starts with one fixed axis (fertility), and students perusing other rows of data and looking for co-variances that might explain the development of fertility rates. They might, for example, discover that a negative development of fertility rates is closely dependent on female education, on the availability of clean water or health care, or the GDP. These indicators are probably easily seen as part of the problem description, as they fit into current models of demographic change taught at school and university levels.

Things become more difficult as soon as students look into co-variances with data either not obviously linked to the phenomenon to be explained, or clearly unlinked to it, but still co-variating. In practice, students mentioned, as an example, fertility co-variating with CO_2 emissions: the higher the CO_2 emissions, the lower the fertility rate – a seeming correlation that needs further explanation (see, for example, www.bit.ly/UH0xXL).

Things may become even more complicated for teachers when data sets co-variate that almost definitely have no direct links, such as fertility and Internet use (see Fig. 15-4). However, it is exactly this type of data that allows students to become aware of their own hypothesis-generation. When asked for possible explanations of the above seeming link, they came up with the following hypotheses:

- 'More highly developed countries can afford the Internet more easily, even in small households'
- 'More Internet use means more knowledge about contraception'
- 'More Internet use leads to staying at home, reducing social contacts'
- 'Having the laptop on your legs is not healthy, leading to infertility'
- 'Because more people use the Internet, they have less time to interact sexually'
- 'It's porn movies instead of sex'

(All examples taken from 4 GIS-day groups, 16.11.2011)

Guiding questions to evaluate hypotheses developed by students might be:

- Does the hypothesis use data that can be considered indicators for something completely different (i.e., links between Internet and education, or Internet availability and wealth)?
- Does the hypothesis have a coherent structure? Does it, for example refer to scientific evidence for the effects on health of Internet use?
- Does the hypothesis take into account human agency as the starting point of human behaviour?
- Does the hypothesis fit into existing theories, or does it break new ground, and therefore possibly lead to innovation?
- Does the hypothesis refer to lay knowledge to explain the situation?

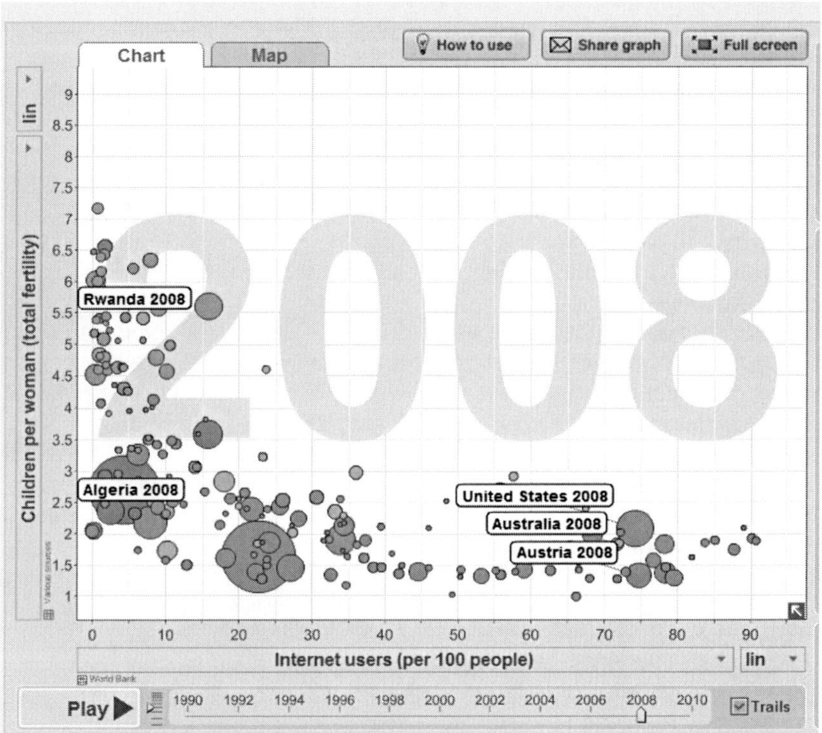

Figure 15-4: Fertility and Internet Usage.

These questions may in many cases be answered by doing another round of research using the databases available. It is for teacher and learners to determine when the exploration is complete.

The visualizations generated can also be used by students to support their argument. In this case, the aim of the exercise is to make students aware of the potentially persuasive effects on discursive processes of using visualization, and in reasoning with and convincing their audience through the image they have developed. The tasks then could be: *Develop your own hypotheses, and convince your classmates of its truth using the best possible visualization you can generate.*

The task could include sub-tasks like: *Describe the rationale of your hypothesis; justify your specific selection of design and visualization tools for your argument.*

3 Preparation for the Teacher

The learning environments described above have been tested with 4 different student groups aged approximately 16 years old during a GIS-day event. They have also been used in various undergraduate teacher training and in-service teacher training courses. Both cases together were originally presented in one 45-minute session. However, to go deeper into both exercises, division into two units seems preferable. More complex questions, for example looking for data sets that explain other data sets, may take longer.

The materials required are ubiquitous: technically, these are limited to a standard PC and an Internet connection. Gapminder is available free of charge, as are all the data associated with it. For exercise (a), the material and procedure presented here are sufficient. The material requirements are a PC, Internet connection and a digital projector, a large sheet of preferably white paper, and coloured stickers for the individual estimates. The example may be linked back to the students' home country (not all countries have this typical demographic change; teachers need to be aware of local data in advance). Strict assessment of the results is not appropriate here, as the point is to lead students to an awareness of the vagueness of estimates.

Case (b) is a little more complex. Technical requirements here are a standard PC with Internet connection for each student or group of 2; a PC for demonstration purposes is useful. The teacher needs to have a specific phenomenon to be explained, be aware of data available within Gapminder, and probably have a few typical hypotheses ready to show students how they might be constructed. The teacher also needs to know how to change between different data sets and, for the last exercise, needs to have some idea of which basic visualization strategies are more convincing than others. It pays to play around with the Gapminder platform for half a day or so, which will be a good investment of time as it can be used for a huge variety of topics (see Lang, 2012).

In all cases, however, learning and the knowledge-production process are exploratory, and reasonable freedom needs to be given to students both to construct their hypotheses and to produce convincing visualizations. Assessment can take different forms. It could relate to the questions evaluating hypotheses given above. Alternatively, the students themselves could assess which they think are the most convincing hypotheses and visualizations presented by their peers. Students can also be asked to present their selections in the hypothesis-generation process as well as the specific use of visualization tools.

Last but not least, with case (b) comes a need for openness to the unexpected, confounding and confusing. The teacher is well advised not to diminish complexity and transfer 'truths' of whatever kind just to make results assessible.

4 Experiences, Alternative Tools and Transferability

In various educational settings, Gapminder has proved helpful across a range of aspects of spatial citizenship (Gryl & Jekel, 2012); the hypothesis-generation can be linked to methodology; case (a) can be linked to reflection on the use of geomedia and their reception; the visualization can be linked to participation and argumentation.

We suggest the following very practical guidelines when working with students at either secondary or post-secondary level:

The reduction to 3 variables (data 1, data 2, time) is usually complex enough to make people think critically; in our experience, more data sets visualized at the same time did not help explanation, but they did make for 'convincing' argumentation support.

When looking for data sets to explain other data sets describing a process, we found it necessary to fix one axis (for example, in our case, keep fertility stable and look for data sets explaining the change in fertility). Keeping both axes open (as you would with a task like 'explain demographic change') was found confusing by users, irrespective of their educational level, at least in short-term projects.

Teachers need to be open to very confounding hypotheses, and probably need specific education as to which hypothesis just describes a coincidence, which data set is an indicator for something completely different, and how to decide on coherent and non-coherent hypotheses.

The general ideas of a hypothesis-generating approach using Gapminder are widely transferable, for example to find out about presumed links between greenhouse gases and economic development, or education and reproduction (see Lang, 2012; Jekel, 2010). Another similar platform is available, albeit with partially different data, in the form of Google's public data explorer (www.google.com/publicdata). Gapminder, here, has the advantage of providing ample material and ideas for teachers. Although the claim that the tool is the basis of a 'fact-based world view' sounds far-fetched (and 'hypotheses-based world view' would be a better description), the toolkit can be recommended for a broad variety of uses in secondary education.

References

Crampton, J. W. (2001): 'Maps as Social Constructions: Power, Communication, and Visualization'. *Progress in Human Geography* 25, 235–252.

Gryl, I. & Jekel, T. (2012): 'Re-Centering GI in Secondary Education. Towards a Spatial Citizenship Approach'. *Cartographica* 47 (1), 18–28.

Jekel, T. (2010): 'Internetnutzung und Fertilitätsrate, oder: Dynamische Visualisierung als Unterstützung eines problemorientierten Unterrichts.' *GW-Unterricht* 118, 38–46.

Lang, R. (2012): 'Using Gapminder.' *GW-Unterricht* 126, 76–87.

CHAPTER SIXTEEN

USING GEOMEDIA FOR COLLABORATIVE LEARNING ENVIRONMENTS: THE EXAMPLE OF PARTICIPATORY SPATIAL PLANNING

ROBERT VOGLER AND SABINE HENNIG

Abstract

Communication and information exchange is increasingly web2.0-mediated, networked and complex. The use and integration of a spatial reference to information, i.e. geomedia, have been gaining importance. These changes and the potential of spatial representations to contextualize learning content account for an increasing relevance of geomedia in education as well. Based on these developments, 'Spatially-Enabled Learning' makes use of web-based mapping to support interaction and communication in educational contexts via 'Social Geocommunication', linking social media with individual spatial representations. The purpose is to enable learners to be 'produsers' (producer-users) of information with a spatial reference. This is helpful in education and everyday life with regard to Spatial Citizenship, i.e. reflective and participatory practice.

After stating the theoretical backgrounds of Social Geocommunication and Spatially-Enabled Learning, this paper will outline an approach for using these in school. The example presented refers to participatory and collaborative spatial planning.

1 New Concepts Needed: Geovisualization, Society and Learning

Our world is becoming more and more interconnected. This refers to societies as well as individuals, in terms of numerous issues such as living environments, economies and social interactions. Communication and information exchange turn out to be the foundation of a postmodern society. With advances in Information & Communication Technologies (ICT), information exchange, discussions and the way we catch up on news and information have changed considerably. This takes place through computer-mediated communication. Meanwhile, the Internet has gained overwhelming relevance in all areas of people's lives while direct contact is increasingly augmented through – and partly even replaced by media-based communication occurring in social media. This is closely related to web2.0 applications which provide inter-operability and numerous innovative opportunities. Users are allowed to share, publish, interact and collaborate with each other in a participatory social media dialogue as produsers (producers and users) of user-generated content.

With the evolution of the mobile Internet in recent years, digital communication practice and information gathering are also increasingly associated with a spatial reference. In consequence, the use and production of geomedia play a pivotal role in web2.0-mediated communication. Due to the appearance of manifold devices (GPS-enabled mobile devices, such as smartphones), applications (digital globes, web mapping tools etc.), services (location-based services etc.), as well as open data (initiatives such as spatial data infrastructures (SDI), open governmental data, the public's interest in and use of geomedia has increased rapidly (Thielmann et al., 2011). Meanwhile, spatial data and geographic applications on the Internet are not restricted to the work of experts; they are accessible and usable by the general public with no competences with regard to spatial data, GIS or classical cartography. In sum, we can say that a spatially-enabled society is being created.

On the premise that school education should provide relevant skills to participate in society, these changes in our everyday life require a stronger integration of geomedia in education. The embedding of geomedia into everyday life requires the development of more sophisticated communication skills in terms of geocomunication. Although there have been several attempts to integrate GI skills in geography education, there have been no recent attempts to support communication in learning processes with a spatial contextualization.

Spatially-Enabled Learning. A paradigm shift from classical technology-based geomedia teaching environments to an approach which uses geomedia to support communication processes in learning scenarios can fill this gap exactly (a shift from 'geomedia in education' to 'geomedia for education'. In other words, this gap can be addressed through Spatially-Enable Learning, an approach which aims to integrate communication and geo-technological tools in a state-of-the-art learning environment. The Spatially-Enabled Learning approach makes use of web-based spatial representations, i.e. geomedia, to support communication processes in educational settings by connecting complex learning content to a spatial dimension. This is achieved by contextualization, by cartographic support of arguments in discussion processes, or simply by a spatialized portfolio that may be individual or collective or both.

The rationale of this approach is based on the idea that spatial representations create a special type of dual coding, corresponding to 'Dual Coding Theory' (Paivio, 1986). The primary notion of this theory is that human cognition is separated into verbal and nonverbal subsystems. Each of these has a corresponding coding system to translate information: analogue codes for abstract verbal information (text), and direct symbolic codes for nonverbal information (images). Empirical research has shown that the response time and the memorability regarding information presented are best when analogue codes and symbolic codes are processed simultaneously. Following this predication for Spatially-Enabled Learning, there is good reason to believe that the link between analogue codes (in our case: learning content) and symbolic codes (in our case: spatial representations) enhances the learner's memory for complex and abstract information. Beyond the fact that visually prepared content in general shows a much greater memorability than verbal or textual content (Weidenmann, 2006), geovisualization and maps especially show high potential to support both the subjective appropriation of space and a collaborative production of meaning (Hennig, Vogler & Jekel, 2011). This can also be useful for the support of learning processes via the spatial contextualization of learning content.

The idea of Spatially-Enabled Learning goes beyond the spatial-thinking approach discussed recently (NRC, 2006) with regard to professional skills in GIS education; it focuses on spatial analysis in the context of absolute concepts of space. Spatially-Enabled Learning uses a constructivist understanding of space, namely that meaning is always attached to space via communication. In this context, spatial representations (like maps or geomedia in general) are very powerful instruments to consolidate such socially-constructed spatial meanings (Lefebvre, 1993). Accordingly, spatially-referenced meaning (in this case: learning content with a spatial reference) can be highlighted and supported with a spatial contextualization, via spatial representations created by learners and/or teachers. This is believed to be helpful in both science education and everyday life with regard to Spatial Citizenship.

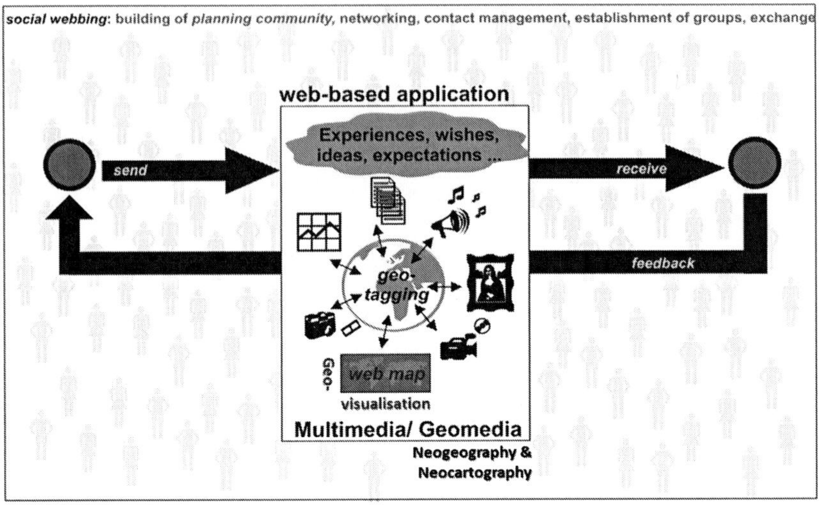

Figure 16-1: The Framework of Social Geocommunication (Hennig, Vogler and Jekel, 2011).

Social Geocommunication. The idea to support communication by using geomedia derives from the concept of Social Geocommunication – linking social media and web2.0 practices with individual spatial representations in participatory spatial planning processes (Hennig, Vogler & Jekel, 2011). Geocommunication has been defined as the use and combination of a multitude of available multimedia and geomedia by high numbers of participants on the one hand, and features enabling user interaction on the other. Besides producing and integrating geomedia and multimedia, Social Geocommunication also leverages aspects of social media served by web2.0 platforms, such as community-building, direct and indirect exchange, and networking. This encourages users to participate interactively in communication processes (discuss, exchange, or simply inform themselves) via (social) web2.0-based tools. Social Geocommunication consequently enables well-structured interaction, consultation and decision-making between all participants by making use of individually produced spatial representations, i.e. digital maps, to explain their own ideas to other people (see Fig. 16-1).

In order to incorporate Spatially-Enabled Learning and Social Geocommunication in education, suitable tools need to be found or created, and subsequently reviewed in order to support the integration of these processes.

2 Teaching Unit: Collaborative Spatial Planning

Shrinking Technical Challenges, or: 'Choose your Tools!'. While in the past, these ideas could not be implemented in education due to the technical challenges involved and the professional cartography skills needed, the development of ICT-based Social Geocommunication allows for some promising new opportunities. A benefit for this is the wide range of web mapping tools freely available online. As these tools are very intuitive to use, there is no need for professional cartography or even GIS skills to be developed. These tools, suitable for applications in learning and teaching, even combine mapping practices with the world of social networking. Different categories of tools can be distinguished (see Table 16-1 for a short review):

(1) stand-alone web mapping applications (web mapping; cartographic communication)
(2) collaborative web mapping applications (co-operative/ participatory web mapping; geocommunication)
(3) social web mapping applications (co-operative/participatory web mapping; Social Geocommunication)

<table>
<tr><th colspan="2"></th><th>web mapping and cartographic communication</th><th>collaborative web mapping and geocommunication</th><th>collaborative web mapping and Social Geo-communication</th></tr>
<tr><td colspan="2">exemplary tool</td><td>ArcGIS online explorer</td><td>Scribble Maps</td><td>TripLine</td></tr>
<tr><td rowspan="2">web mapping</td><td>viewing</td><td colspan="3" rowspan="2">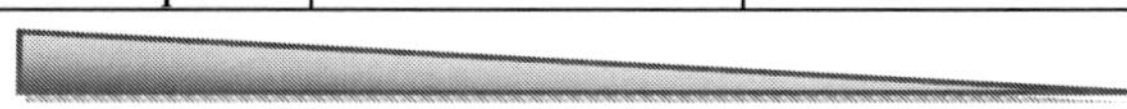</td></tr>
<tr><td>editing</td></tr>
<tr><td colspan="2">social media services and web. 2.0 integration</td><td colspan="3"></td></tr>
</table>

Table 16-1: Web mapping tools in terms of Social Geocommunication opportunities (September 2013)

Applications grouped under these three categories are characterized by a number of different functionalities – implemented to varying extents – to enable web mapping and communication (social networking services: constructing profile, building and maintaining social networks and relations, creating groups, sharing information etc.).

In addition to the examples in Table 16-1, a wide range of other comparable tools cover the pre-requisites of Social Geocommunication and therefore of Spatially-Enabled Learning. The problem is no longer the availability and usability of suitable applications, but more that of identifying them and applying them in an educational context. Even if the chosen mapping tool does not cover all pre-requisites regarding communication services, maps produced by individuals can be embedded in social media or communication platforms. In this case, learners would produce their own maps with a suitable mapping tool and embed the map object in the chosen social media or communication platform (an online forum, a blog, the school Moodle, or even Facebook). Admittedly, the lower the number of participants, the lower the importance of an online communication platform. But even if just a few participants can (perhaps better) communicate face to face, the linkage to a web2.0 platform is highly desirable as it has the advantage of storing the communication processes and, in addition, can serve as a content management system.

The teaching example below, for instance, focuses on the use of Scribble Maps (www.scribblemaps.com; see Fig. 16-2). But this does not imply that the example is restricted to Scribble Maps. If a teacher feels more comfortable with another tool that they have already used, they should feel free to use that one. Scribble Maps has many advantages over other simple mapping tools (e.g. Google Maps or Google Earth): (a) it is easy to use; (b) maps can be produced (collaboratively) and stored online, even without a personal user account; (c) maps produced can be shared or exported to several web2.0 communication platforms. Based on the possibilities it offers for map versioning, map sharing or jpeg-export, Scribble Maps is considered highly suitable for the following teaching example. Besides standard navigation options (pan and zoom), different base maps (Google, ESRI, OSM and others) and features for map creation (showing markers via addresses; events; integrating complex maps with markers, lines or polygons; including multimedia info windows), several more features, e.g. language selection, are available.

We would strongly encourage teachers who are not familiar with Scribble Maps to experiment with it as a user-friendly tool, even if they have no experience of geomedia or digital cartography. Considerable assistance, if required, is available online.

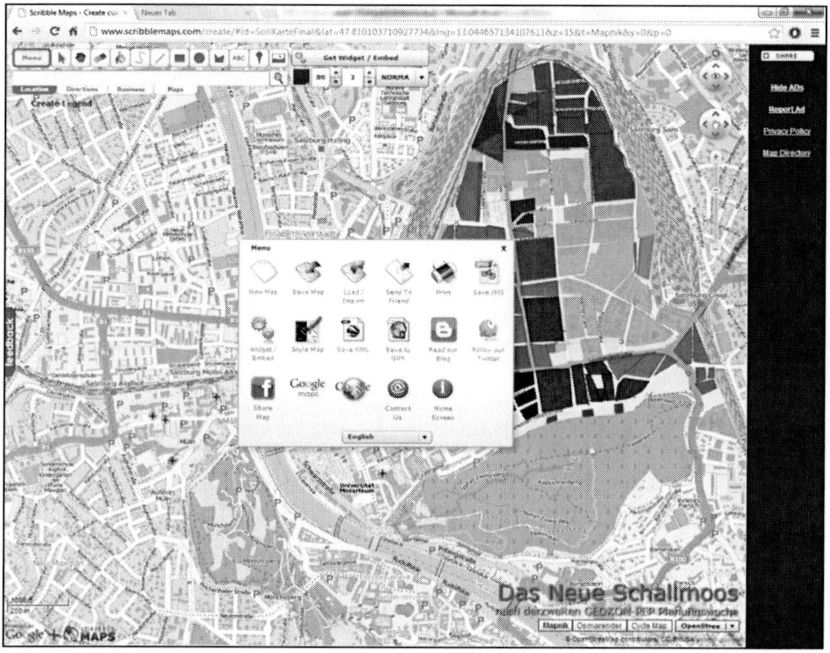

Figure 16-2: The interface of ScribbleMaps, including the toolbar (upper left corner) and the opened main menu (with several exporting/sharing/embedding options), showing a map produced by 16-year-olds in the GEOKOM-PEP project.

Teaching Goals, Background and Target Group(s). The overall goal of this teaching example is that pupils develop collaboratively a spatial planning vision of a specific area while using geomedia to support communication during this process. It can be used in secondary education, with various age groups – the planning object and the complexity of the process can be tailored to students' abilities and are not limited to a specific type of school type.

The idea derives from GEOKOM-PEP (Geovisualization and communication in participatory decision-making processes), an Austrian research project that aimed at participatory spatial planning processes. A communication platform prototype integrating web mapping tools and social networking services was developed to generate 'value-added' geocommunication among pupils and teachers, making use of geovisualization. This platform supported collaborative decision-making in spatial planning scenarios (Hennig, Vogler & Jekel, 2011). Implementing the plan can be highly fruitful in education as well:

collaborative planning processes and constructivist learning scenarios are very similar, as both focus on the collective production of meaning.

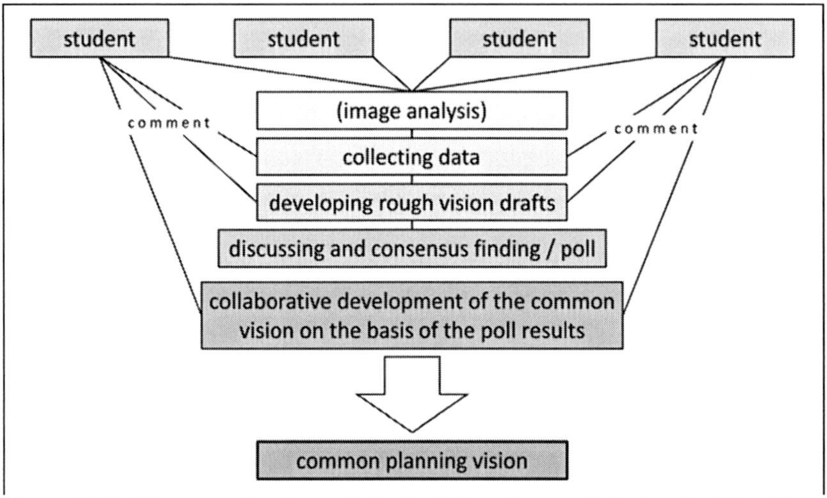

Figure 16-3: The core structure of the 'collaborative spatial planning' teaching example

By keeping the core structure of this one-week workshop (Fig. 16-3) and adjusting it with regard to the planning object, the complexity and the time period available, this teaching example can be also adapted to everyday teaching in everyday school life. This allows for broad possibilities of customization: (1) the planning object can range from the school yard to a whole city district; (2) the time required varies according to the scope and complexity of the planning purpose; (3) the aim is to solve a real-world problem, or a problem conceptualized as a simulation game (comparable to role-playing approaches in education); (4) tools can be chosen as available and suitable.

Step-by-Step Guide. The guide given here is intentionally rough in outline to assist with transferability to other lessons.

working step	**content**	**media, methods, comments**
prerequisites	- technical: • computer lab • availability of an online communication or E-learning platform - abilities: - your students (and you) should be familiar with the use of ScribbleMaps	- projector, PCs with up-to-date browser and flash player, Internet connection - Moodle, an online discussion forum, or a non-public Facebook group
step 0 **...preparation**	- Define the planning object (either prescribe it or negotiate it with your students) - Specify the corresponding planning categories. These categories should be developed in conjunction with your students.	- scale and complexity (school yard vs. city district) depend on the time you want to spend on the entire exercise - depends on the planning object (e.g. traffic, residential areas, green areas, etc. for planning a city district; trendiness, functionality, safety, etc. for planning a school yard)
step 1 **...collecting data**	- Divide your students into small groups *Every group:* - collects necessary data about the recent status of the area - visualizes the recent status with a map using ScribbleMaps, and uploads this map to the chosen communication platform - identifies recent problems regarding the planning categories and posts a short problem report to the chosen communication platform	- 3–4 students per group - group work (depending on the planning object: fieldwork, if needed) - group work (ScribbleMaps) - group work (discussion, short text)

step 2 **…developing particular solutions and reviewing them**	- Every group discusses and develops a first rough planning vision to improve the recent situation. - They then visualize these visions with a map using ScribbleMaps, and verbalize their ideas rougly. They upload both text and map to the chosen communication platform. - Now every group reviews the visions developed by the others and leaves short comments on the platform.	- group work (discussion) - group work (ScribbleMaps, short text) - Every student can now widen his/her point of view and get the bigger picture regarding the problems identified and potential solutions offered by the others.
step 3 **…discussing the collected solutions**	- Define new working groups. - Each group focuses on one of the planning categories defined in step 0 and develops an advanced planning vision based on the demands of their specific planning category. - They visualize these advanced planning visions in a detailed map using ScribbleMaps and upload it to the chosen communication platform. - Each group presents its vision (with the help of the map generated) to the audience. - When every group has presented their vision, the teacher initiates a discussion, where the students try to convince each other with their visions and arguments. -	- 3–4 students per group - group work (results should be based on the drafts developed and stored online in step 2) - group work (ScribbleMaps) - presentations (the teacher should identify potential areas of conflict between the different visions presented) - group discussion (moderated by the teacher)

step 4 **…finding consensus and coming to a common vision**	- When the discussion is finished, the teacher summarizes the issues and outlines the main conflicts, which are transferred to yes/no questions. - The questions are released to the audience for a democratic vote. - The result of the vote leads to a list of democratically legitimated demands with regard to the common planning vision.	- e.g.: 'Do we want to have a playground in our school grounds?' - the teacher notes the outcomes - the document should be uploaded to the platform
post-processing **… the planning results and compiling a planning report** **(optional, but highly recommended to conclude the whole process)**	- Set up new working groups. - Each group generates one part for the final planning report (1. recent status text incl. problems identified; 2. recent status map; 3. vision text; 4. vision map). - When finished, each group presents its results; these (if there are objections from the audience) can be adjusted afterwards. - After completing this step, the 4 parts are put together in one final document.	- 4 working groups - group work (based on the list developed in step 4 and the information gathered and stored online in steps 1–3); the maps will be generated with ScribbleMaps - short (!) presentations - collaboratively developed final planning report as result of the entire exercise.

Table 16-2: Step-by-step guide for the collaborative spatial planning / teaching example

3 Conclusion

The result of the entire exercise (depending on the scope of the study, which can be more or less time-intensive) is a common spatial planning vision for the chosen area. The development of this vision, throughout all the working steps, is based on democratic principles and on a participative-collaborative working pattern following a compromise-based approach. The end product will be an entire planning report.

Because first empirical results show a structuring impact of geomedia use on communication and discussion structures (Hennig, Vogler & Jekel, 2011), the function of geomedia in this teaching example is to help to communicate on spatial patterns; geomedia is not the actual goal of teaching. This resonates with the principles of Spatially-Enabled Learning and is compatible with the main purpose of Spatial Citizenship.

As already stated, a huge range of variations is possible when using this teaching approach. These variations concern not just the planning object, but also the complexity or the time schedule. In addition, some of the four steps can be adjusted or even omitted. Using this open approach, keeping the main structure and adapting it to a chosen object, offers several advantages for the learning goals of the students:

- They are sensitized to the spatial patterns of the planning object and acquire new views on (maybe well-known) areas. (geography education)
- They see that spatial planning is always a question of different interests and that there is never just one correct solution. (geography and political education)
- They see how conflicts can be resolved using step-by-step solutions based on democracy. (political education)
- They develop their discussion, argumentation and empathy skills. (argumentation skills)
- Last but not least, they learn to use geomedia to support communication and argumentation. They see that spatial representations, i.e. maps, can be powerful instruments but are always a subjective construction of meaning.

Based on the experiences of the initial GEOKOM-PEP planning week in 2011, where this teaching approach was piloted with two 11th grade school classes (16 to 17 years of age), it can be highlighted that the use of geomedia shows great potential to structure communication and discussion processes in learning environments. Finally it is important to state that the primary goal or intention of this teaching approach is not learning to use geomedia (although this will happen anyway), but using geomedia to learn.

References

Hennig, S., Vogler, R. & Jekel, T. (2011): 'Web-2.0 Anwendungen zur partizipativen Planung und Sozialen Geokommunikation'. *GIS. Science. Die Zeitschrift für Geoinformatik* 3, 65–74.

Lefebvre, H. (1993): *The Production of Space.* Oxford: Blackwell.

Paivio, A. (1986): *Mental Representations: A Dual Coding Approach.* Oxford: Oxford University Press.

Thielmann, T., Van Der Velden, L., Fischer, F. & Vogler, R. (2011): 'Dwelling in the Web: Towards a Googlization of Space.' *HIIG Discussion Paper Series No. 2012, 03.* Berlin.

Weidenmann, B. (2006): 'Lernen mit Medien'. In A. Krapp & B. Weidenmann (eds.), *Pädagogische Psychologie. Ein Lehrbuch.* 5th edition. Weinheim: Beltz. 423–476.

CHAPTER SEVENTEEN

PLAYING WITH GEOMEDIA TO UNDERSTAND THE COMPLEXITY OF THE WORLD

ERIC SANCHEZ, NICOLAS KRAMAR AND CHRISTELLE LISON

Abstract

The purpose of this chapter is to emphasize the role of geomedia for designing a playful learning situation. We describe Clim@ction, a French-Canadian online role-play game about land use management. Clim@ction involves the students in collaboration and competition in a project designed for implementing green energies in a local territory. The learning situation offers secondary students a complex and non-deterministic problem. Therefore, it is expected that they will develop a broad and realistic view regarding the issue of sustainable development. We describe the background of our project related to personal epistemology and to Game-Based Learning. We also underline the use of geomedia for designing the game, and we present the results of an experiment carried out by teachers and researchers in France and Québec.

1 Introduction

As society changes, schools have no option but to change accordingly. The challenge is to renew education in order to better prepare students and the citizens of tomorrow for the many changes they will face. The Modern period that began in western countries during the Renaissance is characterized by a dramatic change in the relationship to knowledge, with the use of reason as a way to gain freedom, versus arbitrary authorities and traditions. One of the main consequences of this paradigm shift is the scientific revolution of the 17th century that led to a massive development of knowledge in natural sciences and technologies. Despite those developments, in the 20th century the relationship between science and society led to a crisis about problems where science cannot provide definitive solutions or, worse, is considered to be the cause of the problems, as, for example, in the case of the atomic bomb. Today, people are often exposed to socio-scientific issues for which no simple solution can be put forward. Morin (2000) proposed that this new situation should be taken into account by contemporary education. According to Morin, this situation is a consequence of modern ways of thinking that separate scientific disciplines from each other and divide fields of knowledge. Therefore, schools also have to prepare students to manage a great quantity of evolving and strongly interrelated and controversial information.

In this chapter, we describe a process for designing a digital role-play game to enable students to face a complex problem and to deal with controversial information. Our approach is based on the use of geomedia, which enables the design of an authentic and non-deterministic learning situation close to the situation faced by local authorities, companies and citizens when they must make decisions regarding land-use management.

2 Aims: Designing Authentic and Non-Deterministic Learning Situations to Foster Students' Personal Epistemology Development

In contrast to the thought of modern times which divides knowledge and builds knowledge according to the reductionist paradigm, Morin highlights complexity in the following terms:

> Complexus means that which is woven together. [...] In fact there is complexity whenever the various elements (economic, political, sociological, psychological, emotional, mythological ...) that compose a whole are inseparable, and there is inter-retroactive, interactive, interdependent tissue between the subject of knowledge and its context, the parts and the whole, the whole and the parts, the parts amongst themselves. (Morin, 2001, p. 15)

For Morin, a complex topic must be distinguished from a complicated one. For instance, the way 20 balls will interact on a pool table is complicated to calculate, but this problem has one and only one right solution. On the contrary, a complex problem has many possible solutions. Solving a complex problem usually involves dealing with knowledge from very different disciplines, such as humanities and natural sciences, and so a deterministic solution cannot be proposed. From a pedagogical point of view, the example from the game of pool is called a 'well-structured problem'. It differs from an 'ill-structured problem' (King & Kitchner, 1994), which is a complex type of problem for which a consensus between experts cannot necessarily found.

An ill-structured problem relates to a situation that is non-deterministic. There is no single solution, and a range of solutions need to be considered, taking into account both the available information and the identified lack of knowledge.

The ability to manage ill-structured problems is related in part to the relationship an individual has with knowledge and his/her beliefs about it (Jonassen, 2000). More precisely, how an individual considers the nature of knowledge and the nature of knowing will influence the way the individual manages ill-structured problems (King & Kitchener, 1994). This particular type of problem management has been modelled in developmental psychology with the concept of personal epistemology (Hofer & Pintrich, 2002).

In order to characterize the relationship of the students with the knowledge from an epistemic point of view, we referred to the widely accepted conceptual model of Hofer and Pintrich (1997) which characterizes the individual's personal epistemology by way of four types of beliefs. All four beliefs, called dimensions by Hofer and Pintrich, are interrelated in a coherent way, and each one of them can develop on a continuum of representations. The first two dimensions refer to the nature of knowledge. The dimension of 'certainty of knowledge' is related to the idea of truth in knowledge, from the idea that knowledge is well known and doesn't change with time to the idea that it is always evolving. The dimension of 'simplicity of knowledge' expresses the degree to which knowledge is an isolated fact or is more integrated to other knowledge. Because of this particular characteristic, this dimension is strongly related to Morin's definition of complexity. The other two dimensions related to the nature of knowing are 'source of knowledge' and 'justification for knowing'. The source of knowledge ranges from the idea that knowledge resides outside of the self and is transmitted by others, to the idea that the individual is a knower who participates in knowledge construction in interaction with others. Finally, justification for knowing describes how the individual evaluates knowledge claims, ranging from accepting arguments simply on the basis of the the authority of the person expressing the argument, to the acceptance of multiple opinions.

Within this context, Digital Play-Based Learning can be considered an alternative pedagogy to promote the resolution of complex and interdisciplinary problems. Indeed, it has been demonstrated that Digital Play-Based Learning approaches have the power to foster students' motivation to solve complex problems; these possibly more playful learning situations make up spaces of reflexivity where students can test their own ways of thinking and acting by designing and trying out their own solutions. In what follows, we will use the expression 'digital epistemic game' for such a situation.

Our work consists of designing a playful situation, or in this case a game, as a meta-activity, a model of an ordinary situation for sustainable development education. Geomedia are widely used for such an approach. Four criteria distinguish a game from an ordinary learning situation:

- The consequences of the decisions are minimized. Mistakes and failures are permitted and it is always possible to try again.
- The player is allowed to take decisions and the results of the game depend on these decisions, as a result of which:
- Uncertainty always remains with regards to the ending. This ending has not been written in advance by the teacher. In addition, this freedom to take decisions offered to the learner/player is limited by rules:
- Rules are norms and conventions shared by the participants.

This set of criteria is valuable in implementing Game-Based Learning situations in the classroom and to distinguish such situations from Problem-Based Learning and Project-Based Learning, which are different though closely related educational approaches.

From this starting point, our aim was to design Clim@ction, an online multiplayer game, which consists of a complex and authentic situation regarding the issue of 'green' energy production and consumption for local territories.

3 Clim@ction, a Digital Role-Play Game Designed with Geomedia

The design of Clim@ction is based on a collaborative (practitioners and researchers) French-Canadian research project funded by the Social Sciences and Humanities Research Council of Canada. *Clim@ction* is a digital epistemic game about energy and land-use management. We designed this online multiplayer game to offer French and Canadian students the opportunity to collaborate in the design of a project for local energy production and consumption that must propose an innovative and acceptable solution regarding the environmental, human and economic impacts. In this role-playing game, 'international companies' which are specialists in green energy compete to convince 'local authorities' and an 'association of citizens' that they have the best project for energy production.

As a first step, the students use an online platform (adapted from Moodle, a free Learning Management System) to obtain information regarding the availability of the energy sources (wood, water or wind), the place where energy is consumed (residential area, industry ...), and constraints related to land-use management. The students use geomedia, such as Google Maps or databases, that provide data which allow them to identify challenges and opportunities for implementing green energies. For example, they locate the main sites for energy consumption; they recognize local sources of green energies, such as sustainable forestry, solar energy, wind, or wave and tides. Students also get an idea of the difficulties that will be faced in terms of the economic, societal and environmental impacts of the solutions adopted within the different territories.

All the useful information regarding energy and land-use management is available on the online platform. It is authentic because data are provided mainly by real experts from local administrations, before being adapted by teachers to the level of their students. This body of information allows students to gain a complex and realistic view of the issue.

As a second step, the students design a pre-project. Geomedia are used by the students to collaborate and to share ideas and arguments. For example, Google Sketchup is widely used to geolocalize places for energy production (e.g. wind turbines, or methane collection from biomass), and 3D models are integrated into a digital globe in order to create a representation of the consequences of the project. To help students to facilitate meetings, organize collaboration, and obtain grants that will reduce their costs, a panel of experts, including teachers, is in place. This panel also provides help on request and is accessible via an online forum. By using geomedia, the learners/players are involved in the whole process for designing a project; they also have to face issues regarding geo-information processing, such as the availability and reliability of data, and the relevance of data processing.

During face-to-face sessions, the students present their projects and provide arguments or answers to criticisms from the 'local authorities', the 'critical citizens' or the other 'companies'. Geomedia are widely used by the learners/players to design documents that aim at sharing and challenging ideas with partners. They are also used to design an overall representation of their project and in convincing the 'local authorities' or the 'association of citizens'. The learner/player also creates maps and figures that are used during the oral presentation.

After the face-to-face presentations, the students re-design their projects according to the feedback that they receive, endeavouring to take on board those suggestions or criticisms (mainly economic or societal) that seem most important. They aim to create more convincing presentations, and geomedia are again of great importance in supporting this task.

In addition, the 'local authorities' and the 'citizens' verify the coherence of the different projects by gathering relevant data. Students use an augmented reality game designed using MITAR, a program developed by MIT's Scheller Teacher Education Program lab, and implemented on Pocket PCs. Using MITAR, the learners/players acquire information from short videos from different green energy projects in various places in the city. Students are then able to evaluate the impact of their proposals on the field.

A final vote (open to all the students at the school) allows a decision to be made as to which is the best project in terms of environmental, economic and social impacts. The winning project can combine the proposals of different 'companies'. Therefore the competitive dimension of the game is balanced by the opportunity offered to the learners/players to collaborate and to design a project which benefits from the expertise of different 'companies'.

4 Lessons Learned

We collected quantitative and qualitative data that encompassed

(1) the messages (379) posted by the students on the platform

(2) the videos made of the students during the presentation of their projects

(3) the recordings of students made during meetings of five focus groups dedicated to discussing the game

These data were discussed during focus-group meetings of teachers and researchers involved in the project.

The data collected on the platform demonstrate that, at the beginning of the game, students consider sustainable development mainly, and very often only, from an ecological perspective. The names chosen for the companies reflect this: *Eco People Power*, *eCology*, *Eco People*. At this stage, the students' messages on the platform express strong ideas about the importance of taking into account the environmental aspects of the issue: 'We must stop deforestation because we have enough industry and we need nothing more;' 'We are a group of citizens defending everything that is green.' Later on, there is an evolution in the ideas expressed by the students. During the course of the game, they become better able to provide arguments that take into account a broader vision of sustainable development. Some messages include the economic dimension ('The goal of our company is to produce wood in large quantities in order to save the planet and for the lowest price!'), or both the societal and the economic dimensions: 'I have some ideas: reduction of unemployment by 15 % in the region, reduction of total contaminated area of 300,000 m^2 by the company and everything associated with it (transport, etc.), economic benefits to the region more than €1 million per year.'

The evolution of the ideas expressed by students was confirmed by the data collected while elaborating the final presentations of the projects, and during the focus groups that were held two months after the end of the game. The data support the conclusion that there are links between the choice for the design of the game (authenticity and complexity of the situation) and the fact that students changed their points of view regarding the nature of sustainable development. This change relates mainly to the dimension 'simplicity of personal epistemology': knowledge regarding sustainable development is recognized to be interrelated and integrated.

In addition, our results suggest that the epistemology embedded in the game is widely accepted and appreciated by most of the students (because it is 'close to real life'). Despite the learning situation provided by the game being very different from a more conventional school learning situation – many students confessed to having felt insecure at the start of the game – the great motivational dimension of the game allowed them to get over this obstacle and to become involved. Indeed, geomedia offer an excellent opportunity to design a learning situation where students have to face the complexity of socio-scientific issues while benefiting from the security offered by a game which minimizes the consequences of the decisions taken.

References

Hofer, B. K. & Pintrich, P. R. (2002): 'Personal Epistemology as a Psychological and Educational Construct: An Introduction'. In eid. (eds.), *Personal Epistemology: The Psychology of Beliefs about Knowledge and Knowing*. Mahwah, NJ: Lawrence Erlbaum. 3–15.

Jonassen, D. (2000): 'Toward a Design Theory of Problem Solving'. *Educational Technology Research and Development* 48N (4), 63–85.

King, P. & Kitchener, K. (1994): *Developing Reflective Judgment: Understanding and Promoting Intellectual Growth and Critical Thinking in Adolescents and Adults.* San Francisco, CA: Jossey-Bass.

Morin, E. (2001): *Seven Complex Lessons in Education for the Future.* Paris: UNESCO.

Sanchez, E. & Jouneau-Sion, C. (2009): 'Playing in the Classroom with a Virtual Globe for Geography Learning'. In T. Jekel, A. Koller & K. Donert (eds.), *Learning with Geoinformation IV*). Heidelberg: Wichmann. 78–86.

Chapter Eighteen

A Four-Step Model of Participative Planning for Schools

Virpi Hirvensalo and Mika Sihvonen

Abstract

This paper introduces a new four-step model which provides a training tool for students to participate actively in changing the meanings of a place in a democratic process, according to the aims of spatial citizenship education. The model can be applied by students from the level of secondary school and upwards. It is suitable for many kinds of GIS or web mapping service applications. It concentrates on the practical idea of learning with GIS from an emancipatory point of view. The model is based loosely on the pedagogical model of progressive inquiry.

1 Aims of the Learning Environment

Co-developing Local Environments. Schools are opening up to the surrounding society, and within the curriculum there is a need for collaboration with external actors. Children and young people are able to learn collaborative practices and create new ideas for planning and negotiating. In schools, these important practices can be learned via cross-disciplinary themes, such as active citizenship, environmental education and media skills. Communication and co-development amongst participants are seen as particularly important ways to proceed in these new approaches (Staffans et al., 2010, 17–18).

Gryl, Jekel and Donert (2010) introduced the concept of spatial citizenship. They press for a more emancipatory approach to citizenship education. Competences that will enable students to participate more actively in society are as follows:

1) deconstructing the spatial information available from various sources,
2) furthering one's own visions of social space by
3) being able to translate and communicate them with the help of GI.

(Gryl, Jekel & Donert, 2010, 2)

These competences will be explored within our model, which aims at teaching students to act as experts within their own surrounding neighbourhood. This is realized by advising students to observe and assess the quality and functionality of their everyday environment. The model guides them to develop their own vision of social space with a view to proposing improvements; GIS or web mapping services and social media tools serve as instruments. Finally, students create map presentations of their visions and share them for the negotiation with other students or members of the local community. The model is applicable to students from secondary school through to higher education. Younger students need more guidance and the support of prepared materials from their teacher, while advanced students are able to work more independently.

Learning by Inquiring about Participative Planning. This four-step model of participative planning is an application of progressive inquiry. It was developed by Paavolainen and Hakkarainen with their research group.

In this model, explanation-seeking processes and collaborative and social aspects of learning are emphasized (Muukkonen, Hakkarainen & Lakkala, 1999; Paavola & Hakkarainen, 2005, 549). Using space as a mediating artefact, students develop common objects of activity, which in this case are proposals to improve the chosen place.

In order to strengthen the students' perception of the identify of place, they practise discerning the characteristics of a place and evaluating them. Following fieldwork, the location is explored with the help of textual and visual materials, which deepens learning. With its physical features and through the meanings associated with it, space as a mediating artefact functions as a support for contextual learning by relating observations, stimuli and experiences to a learning process (Staffans et al., 2011, 120).

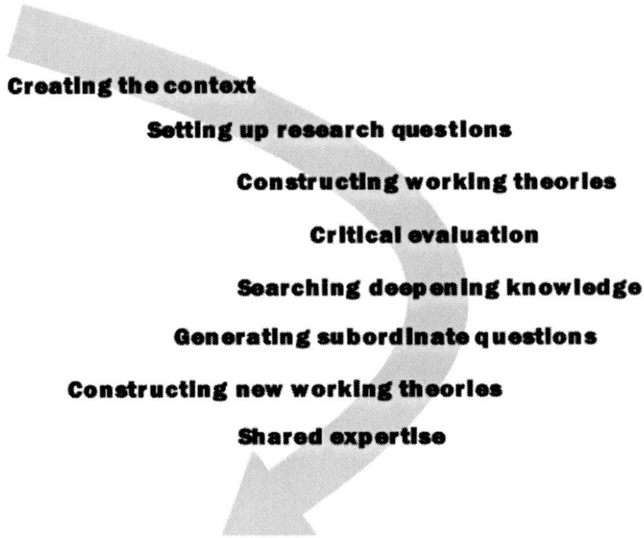

Figure 18-1: Progressive inquiry includes eight phases (modified from Muukkonen et al., 1999)

For students it is also essential to perceive various, possibly conflicting, meanings, which are socially constructed and attached to physical space. Strong place identity and understanding of the diversity of meanings are premises, firstly, for realizing democracy in the planning process and, secondly, for participating in it by creating one's own innovative visions of social space.

Paavola and Hakkarainen (2005, 550) mention the importance of publishing students' results locally. The nature of learning changes when producing knowledge for an authentic audience. With this model, collaborative practices encompass the co-development of improvement proposals with other students and the local community by using tools of geo- and social media. In this way, the local community is seen as a learning community in a wider sense.

2 Description of the Four-Step Model of Participative Planning for Schools

Step 1: Choose the Place and Define Need(s) for Change. The first task is to choose the place for which a proposal for improvement will be created. This is equivalent to the phase called *set up the context* in progressive inquiry. The place can be any size, such as a street, brownfield site or neighbourhood. If a municipality has a planning project under way, this exercise can be connected to it. Students acquaint themselves with the chosen location by acquiring information about it in small groups.

The second task is to outline the need(s) for change and present them as problems. This corresponds to *presenting a research problem*, which is the second phase in progressive inquiry. Small groups, or a whole class together, can consider the following questions:

- Is there need for small- or large-scale renewal? Which elements are in need of improvement?
- Is there a need to improve the social or natural environment, a built-up or undeveloped environment?
- Does the need for change relate to to the cleanliness, beauty or ecology of the place, or something within the scope of active citizenship?

It is also possible to take take into account the seasons, times of the day, or viewpoints of different groups of city dwellers. At this stage, it is enough to define the problem, for example, *how to make a street more liveable.*

Extra step: create a plan. If the time schedule permits, a plan for the realization of an improvement idea is made. This task is similar to the phases *construct a working theory* and *evaluate it critically* in progressive inquiry. The plan should include the following elements:

- What is done and observed in the field?
- How is the work to be distributed in small groups?
- How is the presentation of an improvement idea realized in practice?
- How will the presentation be promoted, and to whom will it be made?

The small groups comment on each other's plans. It is important to concentrate on constructive feedback, focusing on ideas.

Step 2: Observe the Place. Students explore the place, which covers both the fifth and the sixth phases in progressive inquiry, aiming to deepen their knowledge and generating subordinate questions. Students should move around and observe the chosen location freely, taking photographs, filming and recording according to what equipment is available to them, and making notes. A GPS is an option to save data in the field and transfer it to a web-based GIS application, or a web map. Below is a list of questions to think about in the field. Additionally, it is important for students to create their own questions, based on their present understanding of the place.

- What are your feelings about the place or other locations (good, bad, excited …)? What are the reasons for these feelings?
- How easy is it to move around? Are there any problems or obstacles etc. within the chosen location?
- Are there any traffic problems?
- Something missing, under- or over-represented in the place?
- If you had unlimited resources and there were no planning restrictions, what changes would you make ?

Step 3: Create a Proposal for an Improvement. The students' task, in small groups, is to create a proposal in the form of a map presentation, using a GIS application, or a web mapping service. This is equivalent to the phase *constructing new working theories* in progressive inquiry.

If necessary, students can reconsider their proposals by acquiring more information from other sources. Guidelines for the map presentation are as follows:

a) Outline your improvement. Describe your idea in a few sentences.

b) Mark the place for your proposal as a point, line or area object on the shared map. Then add spatial information of your choice. Spatial information can be presented in visual, textual or audio format. In order to present your proposal, you can, for example, combine descriptive texts and images. It is also possible to alter images of the place with a graphics editor, for example by adding objects like a bench to an image of a park.

The practicality and the level of innovation shown by the proposals are equally important. Alternatively, proposals can be based on fantasy. They can be *dream places*, having no relevance in real life.

Figure 18-2: A group of students from Turun Lyseo upper secondary school have described new usages for sports grounds in the Impivaara sports area of the city of Turku, Finland. (Source: PaikkaOppi e-learning environment)

Step 4. Present your Proposal and Vote: Finally, students present their proposals to their classmates, or to the coordinator of the municipal spatial planning project. This task is similar to the last phase in progressive inquiry, *sharing expertise*. It is also possible for students to vote for the most feasible proposal. As a continuation of the exercise, the whole class can promote the winning proposal in diverse social media. In that case, a group of students creates a page, for example in Facebook, with a description of the proposal, and link the map presentation to it. In the best of cases, the proposal can give rise to discussion in the local community about the needs for change at the local level.

3 Teacher's Preparation

Suitable materials for this exercise are maps and images, for example aerial views and street views of the chosen location, which can be found in online map services. Useful reports and plans are often found on municipalities' websites. Inviting a guest – for example a researcher, the coordinator of a planning project, or a local community leader – is an excellent way to learn interesting facts about the place. It is also recommended that students acquaint themselves with documents describing the planning process in general in their own area. For example, in Finland, such documents are found on the web pages of the spatial planning departments of the largest cities.

The whole project is carried out in small groups, which are formed at the outset, and takes 3–5 lessons of 75 minutes each. The step-by-step procedure for the teacher is given below.

Step 1: Choose the Place and Define Need(s) for Change. The first task of this step, 'choosing the place', takes 0.5–1 lesson, and the teaching method is a literature review and / or guest lecture(s). The second task, 'define need(s) for change', takes 0.5 of a lesson. The teaching method here is discussions in small groups.

The aim of this step is to commit and motivate students to study the subject more thoroughly. At this point, students begin to learn the process of building their own meanings related to a physical space, a process that continues until step 3. To make the subject more appealing, the students should be able to choose their topics relatively freely within the framework of a fixed main theme.

The extra step, *Create a plan*, lasts 0.5 of a lesson. The teaching method is group work and commenting on each other's plans. This critical evaluation concerns issues such as weaknesses in plans and how to mitigate them. The aim of this step is to strengthen the students' own thinking and have them learn to negotiate democratically. This step should also help to differentiate between their intuitive ideas and new knowledge (see Muukkonen, Hakkarainen & Lakkala, 1999, 407).

Step 2: Observe the Place: A field trip to the chosen location takes 1 lesson. Prepare students with the instruments that will be needed in the field. Instruments for saving observations can be cameras, tablets, audio recorders, paper and pencil, GPS, or a map of the place (electronic or printed). The aim is for the students to learn to observe and assess the quality and functionality of their everyday environment. Contextual learning with field observations provides deeper understanding of the physical features and meanings of the space than could have been acquired by using literature or maps only.

Step 3: Create a Proposal for an Improvement. This step takes place in the classroom. Students work in small groups with computers. The step takes 1–2 lessons. If necessary, internalize the instructions of a web-based GIS application or web-mapping service before you go on any further. Reserve time for the registration process, and create a shared map application for your course beforehand, if required. The aim of this step is for students to express their own meanings by using GI (cf. Gryl, Jekel & Donert, 2010, 9).

Step 4: Present your Proposal and Vote. Teaching methods are small group presentations, taking 0.5–1 lesson, and promotion of the presentations on the Internet, which also lasts approximately 0.5–1 lesson. Methods for student assessment are interactive map presentations, in which notice should be taken of the feasibility and innovativeness of the proposals. They should be informative from the viewpoint of the target group(s). In the learning process, the focus is on students' collaborative skills to create their own improvement proposals on the basis of information collected from the field and from secondary sources, and then to negotiate regarding their plans. The aim of this step is to learn to discuss, stand up for and renegotiate those meanings of space in order to further one's own interest democratically in a planning process (cf. Gryl, Jekel & Donert, 2010, 9).

4 Lessons Learned

This model was piloted in two upper secondary schools in south-west Finland in autumn 2012. Students participated in the planning process concerning the renewal of the Impivaara sports area in Turku, Finland. Map presentations of the improvement ideas were made with PaikkaOppi, a Finnish Internet-based learning environment for GIS studies. The first test group shared their results using SlideShare and the other via Facebook.

Based on these practical experiences, some observations about the functionality of the model in real life are presented here. The aim of progressive inquiry is to encourage students to act as experts using knowledge-building practices such as social interaction, presenting ideas to other people, and sharing intellectual activity (cf. Muukkonen, Hakkarainen & Lakkala, 1999, 3-4). In this exercise, students learn to act as experts on their own surrounding neighbourhood. Two pilot courses showed that many students had not learned to assess physical space or the meanings attached to it, so for them it was not easy to discuss or invent improvement ideas. Therefore, it is advisable to give students some hints regarding possible viewpoints or titles for proposals, such as *a dream park for families*.

Time management during the course might be challenging. First, technical problems with computers or web services can arise. If the services are not well known to the students, it is recommended to allocate time for familiarization first. In the pilot courses problems occurred, for example with the registration on PaikkaOppi. Secondly, it is recommended that one or more short sessions should be arranged after the course in order to look through the possible comments in social media and to hear the students' thoughts about them.

The presentation of the other pilot group did not stimulate discussion in social media, even though the students' presentation was viewed in SlideShare over a hundred times within a month. In order to promote the discussion of proposals, a clear vision for the use of social media is needed. Popular tools for information sharing such as Facebook and Twitter can be useful (see e.g. Lankshear & Knobel, 2011, 57–58), but challenges may lie in the policies and regulations that different schools have concerning the use of social media. As some teachers are reluctant to use their own accounts or profiles to share the ideas in social media, it cannot be made mandatory for the students to do so either. A Facebook page might be created by the school, or even by representatives for the neighbourhood concerned in the proposal, and the discussion can be placed on those forums and linked to other information-sharing services.

We recommend that the teacher utilizes the students' existing skills in social networking. Sharing and tagging the information are quite similar activities, but the students may have helpful views on how to gain attention and attract more people to participate. Classroom conversations about different social networking tools and their use promote the general level of awareness of modern technology and might help students to choose the right tools for different situations. In the classroom, students are able to learn how to build virtual groups, how to invite participants, how to select topics, how to moderate conversations, and how to embed information and media from other sources.

The fundamental idea behind the use of social networking sites is to demonstrate the basics of citizens' online participation. Students also gain a deeper understanding of how a local community can be active online and of what efforts such activity demands. Setting goals for the use of social networking is essential and good ground for classroom conversations. We suggest that students pay attention to the critical use of data sources. Students are able to learn how to distinguish well-argued online material from shallow comments. If available, we recommend that links to municipal documents should be used, as well as recent blog posts or any relevant public debate. In social media, people tend to gather together with other like-minded individuals. Dialogue between the different perspectives can therefore be problematic. The student group and the teacher can scour online newsgroups and forums for popular initiatives that might be useful in drawing attention to the proposal.

Students' improvement proposals take the form of map presentations realized with a web-based GIS application, or a web mapping service. This approach enables the creation of customized maps by viewing, visualizing and sharing the local data. Examples of such applications are ArcGIS Explorer, Google Maps, or a web-based learning environment for GIS studies such as the Finnish PaikkaOppi. Additionally, it is possible to process images using graphic editors, such as Paint or Gimp.

References

Gryl, I., Jekel, T. & Donert, K. (2010): 'GI and Spatial Citizenship'. in T. Jekel, A. Koller, K. Donert & R. Vogler (eds.), *Learning with Geoinformation V*. Heidelberg: Wichmann. 2–11.

Muukkonen, H., Hakkarainen, K. & Lakkala, M. (1999): 'Collaborative Technology for Facilitating Progressive Inquiry: Future Learning Environment Tools'. In C. Hoadley & J. Roschelle (eds.), *Proceedings of the 1999 Conference on Computer Support for Collaborative Learning (CSCL '99)*). Mahwah, NJ: Lawrence Erlbaum. 406–415.

Lankshear, C. & Knobel, M. (2011): *New Literacies. Changing Knowledge and Classroom Learning* (3rd edition). Buckingham: Open University Press.

Paavola, S. & Hakkarainen, K. (2005): 'The Knowledge Creation Metaphor: An Emergent Epistemological Approach to Learning'. *Science & Education* 14, 535–557.

Staffans, A., Hyvärinen, R., Kangas, M. & Turkko, A. (2010): 'Schools as Learning Environments'. In R. Smeds, L. Krokfors, H. Ruokamo & A. Staffans (eds.), *InnoSchool – Networks of Learning, Environments and Pedagogy* (Simlab Report Series 31). Aalto: Aalto University School of Science and Technology.

CONTRIBUTORS

Thomas R. Baker
Esri and the University of Kansas (USA)
tbaker@esri.com

Sue Bermingham
Manchester Metropolitan University (UK)
s.h.bermingham@mmu.ac.uk

Julian Bette
University of Münster (Germany)
julianbette@googlemail.com

Alexandra Budke
University of Cologne (Germany)
alexandra.budke(at)uni-koeln.de

Cyrille Chopin
Geography Teacher (France)
cyrille.chopin@ac-rouen.fr

Giovanni Donadelli
University of Padua (Italy)
giovanni.donadelli@gmail.com

Florian Fischer
University of Salzburg (Austria)
Florian.Fischer@sbg.ac.at

Sylvain Genevois
University of Cergy-Pontoise (France)
sylvain.genevois@u-cergy.fr

Roland Goetzke
University of Bonn (Germany)
r.goetzke@geographie.uni-bonn.de

Inga Gryl
University of Chemnitz (Germany)
inga.gryl@gmail.com

Sabine Hennig
University of Salzburg (Austria)
sabine.hennig@sbg.ac.at

Virpi Hirvensalo
University of Turku (Finland)
vijoke@utu.fi

Henryk Hodam
University of Bonn (Germany)
h.hodam@geographie.uni-bonn.de

Steffen Höhnle
University of Erlangen/Nürnberg (Germany)
steffen.hoehnle@ewf.uni-erlangen.de

Thomas Jekel
University of Salzburg (Austria)
thomas.jekel@sbg.ac.at

Detlef Kanwischer
Goethe University Frankfurt/Main (Germany)
kanwischer@geo.uni-frankfurt.de

Joseph J. Kerski
Esri and the University of Denver (USA)
jkerski@esri.com

Nicolas Kramar
Musée de la nature/ Musée du Valais, Sion (Switzerland)

Ülle Liiber
University of Tartu (Estonia)
ulle.liiber@ut.ee

Christelle Lison
Université de Sherbrooke (Canada)
pedagogie@USherbrooke.ca

John Lyon
Geographical Association, Sheffield (UK)
jlyon@geography.org.uk

Andreas Rienow
University of Bonn (Germany)
a.rienow@geographie.uni-bonn.de

Lorena Rocca
University of Würzburg (Germany)
lorena.rocca@unipd.it

Jüri Roosaare
University of Tartu (Estonia)
juri.roosaare@ut.ee

Vaike Rootsmaa
Jaan Poska Gymnasium (Estonia)

Eric Sanchez
IFE-ENS Lyon (France)
eric.sanchez@ens-lyon.fr

Jan-Christoph Schubert
University of Münster (Germany)
jcschubert@uni-muenster.de

Mika Sihvonen
University of Tampere (Finland)
mika.sihvonen@uta.fi

Hans-Jörg Stark
University of Applied Sciences and Arts Northwestern (Switzerland)
hansjoerg.stark@fhnw.ch

Jérôme Staub
IFE-ENS Lyon (France)
Jerome.staub@gmail.com

Josef Strobl
University of Salzburg (Austria)
Josef.Strobl@sbg.ac.at

Robert Vogler
University of Salzburg (Austria)
Robert.vogler@sbg.ac.at

Kerstin Voß
University of Education Heidelberg (Germany)
voss@ph-heidelberg.de